Contents

Foreword: Sebeok's Biosemiotic Contribution to Cybersemiotics
Søren Brier. 5

Articles

Marcel Danesi:
Modeling Systems Theory. 7

John Deely:
The Quasi-Error of the External World 25

Kelevi Kull:
Thomas A. Sebeok and Biology 47

Susan Petrilli:
Sebiok's Semiosic Universe and Global Semiotics 61

Augusto Ponzio:
Thomas A. Sebeok's Global Semiotics 80

Obituary

Søren Brier:
Thomas Sebeok: Mister (Bio)semiotics 102

ASC Pages

Søren Brier, Trustee:
The Integration of Second Order Cybernetics, Cognitive
Biology (Autopoiesis), and Biosemiotics 106

Book Review

Marcel Danesi:
Signs of Life. 110

The artist of the issue is Luciano Ponzio

CYBERNETICS & HUMAN KNOWING
A Journal of Second-Order Cybernetics, Autopoiesis & Cyber-Semiotics

Cybernetics and Human Knowing is a quarterly international multi- and transdisciplinary journal focusing on second-order cybernetics and cybersemiotic approaches.

The journal is devoted to the new understandings of the self-organizing processes of information in human knowing that have arisen through the cybernetics of cybernetics, or second order cybernetics its relation and relevance to other interdisciplinary approaches such as C.S. Peirce's semiotics. This new development within the area of knowledge-directed processes is a non-disciplinary approach. Through the concept of self-reference it explores: cognition, communication and languaging in all of its manifestations; our understanding of organization and information in human, artificial and natural systems; and our understanding of understanding within the natural and social sciences, humanities, information and library science, and in social practices like design, education, organization, teaching, therapy, art, management and politics.

Because of the interdisciplinary character articles are written in such a way that people from other domains can understand them. Articles from practitioners will be accepted in a special section. All articles are peer-reviewed.

Subscription Information

For subscription send a check in $US (drawn on US bank) or £UK (drawn on UK bank or Eurocheque), made payable to Imprint Academic to PO Box 1, Thorverton, EX5 5YX, UK, or credit card details (Visa/Mastercard/Amex), including card expiry date. For more information contact Sandra Good. sandra@imprint.co.uk

Price: Individual $63 / £40.50. Institutional: $121 / £78
50% discount on complete runs of back volumes.

Editor: Søren Brier, Royal Veterinary and Agricultural University, Economics and Natural Resources, Section for Learning and Interdisciplinary methods, Copenhagen.
sbr@kvl.dk Phone: +45 35282689, Fax: +45 35283709
www.flec.kvl.dk/personalprofile.asp?id=sbr&p=engelsk

Associate Editor: Jeanette Bopry, Instructional Design & Technology, University of North Dakota, Box 7189, Grand Forks, ND 58202, USA
bopry@und.nodak.edu

Art editor: Bruno Kjær, Royal School of Library and Information Science, Aalborg Branch

Journal homepage: www.imprint-academic.com/C&HK
Full text: www.ingenta.com/journals/browse/imp

Editorial Board

Peter Bøgh Andersen
Dept. for Information and Media Science, Aarhus Univ., Denmark

Evelyne Andreewsky
INSERM-TLNP, France.

M.C. Bateson
George Mason Univ.
Fairfax VA 22030, USA

Dirk Baecker
U. Witten/Herdecke Fakultaet fuer Wirtschaftswissenschaft, Germany

Pille Bunell
LifeWorks Environmental Consulting, Vancouver, Canada

Rafael Capurro
University of Applied Sciences, Stuttgart, Germany

Marcel Danesi
Semiotics and Communication Studies, Toronto U. Canada

Terrence Deacon
Dept. of Anthropology
Boston University, USA

Ranulph Glanville
CybernEthics Research
Southsea, UK

Ernst von Glasersfeld
Amherst, Mass., USA

Jesper Hoffmeyer
Dept. of Biological Chemistry
Univ. of Copenhagen, Denmark

Louis Kauffman
Dept. of Math. Stat. Comp. Sci.
Univ. of Illinois, Chicago, USA

Klaus Krippendorff
School of Communications
University of Pennsylvania, USA

George E. Lasker
School of Computer Science
Univ. of Windsor, Canada

Ervin Laszlo
The General Evolution Group
Montescudaio, Italy

Humberto Maturana
Univ. de Chile, Santiago, Chile

John Mingers
Warwick Business School,
Univ. of Warwick, UK

Edgar Morin
Ecole des Hautes Etudes en Sciences Sociales, Paris, France

Winfred Nöth
Wiss. Zent. f. Kulturforschung
University of Kassel, Germany

Roland Posner
Arbeitsstelle für Semiotik
Technische Universität, Berlin

Lars Qvortrup
Dept. of Interactive Media
University of Southern Denmark

Kjell Samuelson
Stockholm University & Royal Inst. of Technology, Sweden

Bernard Scott
Cranfield University at the Royal Military College of Science, UK

Fred Steier
Interdisciplinary Studies University of South Florida

Robert Vallée
Directeur Général, Org. of Systems and Cybernetics, Paris, France

Consulting editors:

Hanne Albrechtsen
The Royal School of Librarianship, Copenhagen, Denmark

Dan Bar-On
Ben-Gurion Univ. of the Negev, Beer-Sheva, Israel.

Jon Frode Blichfeldt
Work Research Inst., Oslo, Norway

Geoffrey C. Bowker
Graduate School of Lib. & Inf. Sci., Univ. of Illinois, USA

Philippe Caillé
Inst. of Applied Systemic Thinking, Oslo, Norway

Allan Combs
Dept. of Psychology, U. of North Carolina at Asheville & Saybrook Inst., San Fransisco

Elisabeth Davenport
Dept. of Comm. & Inf. Stud. Queen Margaret College, U.K.

David J. Depew
Dept. of Communication Studies U. of Iowa, USA

Anne Marie Dinesen
Center for Semiotic Research Univ. of Aarhus, Denmark

Daniel Dubois,
Inst. de Math. U. de Liege, Liege, Belgium.

J. L. Elohim
Instituto Politecnico, Nacional Mexico City, Mexico

Claus Emmeche
Niels Bohr Inst. Copenhagen, Denmark

Darek Eriksson
Mid Sweden School of Informatics, Ostersund, Sweden

Hugh Gash
St. Patrick's College of Education, Drumconda, Ireland

Felix Geyer, SISWO
Netherlands Univ. Inst. for Coordination of Research in Soc. Sci., Amsterdam, Netherlands

Steen Hildebrandt
Dept. of Organization and Management, The Aarhus School of Business, Aarhus, Denmark

Jixuan Hu
School of Business and Pub. Manag. George Washington U. Washington DC, USA.

David Johnson
Dept. of Philosophy, North Adams State College, USA

Pere Julià
Inst. f. Advanced Studies C.S.I.C., Palma de Mallorca

Dr. Shoshana Keiny
Dept. of Education
Ben-Gurion Univ. of the Negev, Beer Sheva, Israel

Ole Fogh Kirkeby
Dept. of Management, Politics & Philosophy, Copenhagen School of Economics, Denmark

Kalevi Kull
Dept. of Semiotics
Tartu University, Estonia

Marie Larochelle,
Dept. de didactique, de psychopedagogie, Universite Laval

Allena Leonard
Viable Systems International Toronto, Canada

Morley Lipsett
Center for Policy Research on Sci. & Tech., Simon Fraser Univ., Vancouver, Canada

Gerald Midgley
Centre for Systems Studies, Univ. of Hull, U.K.

Asghar Minai
School of Arch. & Plan. Howard Univ., Washington D.C., USA

Andrea Moloney-Schara
Georgetown Family Center Arlington, Virginia, USA

Massimo Negrotti
Univ. Degli Studi Di Urbino IMES, Urbino, Italy

Per Nørgaard
The Royal Academy of Music Aarhus, Denmark

Makiko Okuyama
Ohmiya Child Health Center Toro-Cho, Ohmiya-Shi, Japan

Marcelo Pakmann
Psychiatric Services, Behavioral Health Network, Springfield, MA, USA

Peter Pruzan
Dept. of Management, Copenhagen School of Economics, Denmark

Axel Randrup
Center for Interdisciplinary Studies, Roskilde, Denmark

Yveline Rey
Centre d'Etudes et de Recherches sur l'Approche Systémique Grenoble, France

Robin Robertson
General Editor, Psychological Perspectives, Los Angeles

Wolff-Michael Roth
SNSC, Faculty of Education Univ. of Victoria, Victoria, BC

David Russell, Centre for Research in Social Ecology U.of Western Sydney, Hawksbury, Australia

Stan N. Salthe
Natural Systems, New York

Eric Schwarz
Centre Interfac. D'Études Syst. U. De Neuchâtel, Schweiz

N.Sriskandarajah, Dept. of Economics and Natural Resources, Royal Veterinary and Agricultural U., Copenhagen

Stuart Sovatsky
East-West Psychology Program California Inst. Integral Studies San Francisco, CA, USA

Ole Thyssen
Dept. of Management, Politics & Philosophy, Copenhagen School of Economics, Denmark

Mihaela Ulieru
Dept. of Mechanical Engineering The University of Calgary Calgary, Alberta, Canada

Bruce H. Weber
Dept. of Chemistry
California State University Fullerton, CA, USA

Copyright: It is a condition of acceptance by the editor of a typescript for publication that the publisher automatically acquires the English language copyright of the typescript throughout the world, and that translations explicitly mention *Cybernetics & Human Knowing* as original source.

Book Reviews: Publishers are invited to submit books for review to the Editor.

Instructions to Authors: To facilitate editorial work and to enhance the uniformity of presentation, authors are requested to send a file of the paper to the Editor on e-mail. If the paper is accepted after refereeing then to prepare the contribution in accordance with the stylesheet information on the preceding two pages. Manuscripts will not be returned except for editorial reasons. The language of publication is English. The following information should be provided on the first page: the title, the author's name and full address, a title not exceeding 40 characters including spaces and a summary/ abstract in English not exceeding 200 words. Please use italics for emphasis, quotations, etc. Email to: sbr@kvl.dk

Drawings. Drawings, graphs, figures and tables must be reproducible originals. They should be presented on separate sheets. Authors will be charged if illustrations have to be re-drawn.

Style. CHK has selected the style of the APA (*Publication Manual of the American Psychological Association*, 5th edition) because this style is commonly used by social scientists, cognitive scientists, and educators. The APA website contains information about the correct citation of electronic sources. The APA Publication Manual is available from booksellers. The Editors reserve the right to correct, or to have corrected, non-native English prose, but the authors should not expect this service. The journal has adopted U.S.English usage as its norm (this does not apply to other native users of English).

Accepted WP systems:
MS Word and rtf.

Foreword: Sebeok's biosemiotic contribution to Cybersemiotics

Søren Brier

In December 2001, the founder and editor of *Semiotica,* and originator of biosemiotics died. The present issue, is an homage to Thomas Sebeok. This journal was conceived for the integration of the semiotic perspective with cybernetic and autopoietic viewpoints. Sebeok became one of the advisory editors of this journal to show his support for the introduction the doctrine of biosemiotics into the areas of cybernetics, systems, autopoiesis and information theory.

With his support the interdisciplinary effort of CHK grew to include the investigation of overlapping interests in the study of circular processes of signification and communication in both second order cybernetics/autopoiesis and biosemiotics. The point of departure for each is the life world of the organism. Peircean biosemiotics also includes the observer in its phenomenological view of signification. With the active support of Jesper Hoffmeyer, Claus Emmeche, and Kalevi Kull biosemiotic evolutionary views were introduced, and these contributed to the development of a foundation for cybersemiotics. There are clear overlaps between autopoiesis, second order cybernetics, and Jacob von Uexküll's concepts of Umwelt and Innenwelt. They all build on a self-organized, circular causality view of the organism and its life world seen as a whole.

Modern Peircean biosemiotics is very different from the symbolic semiotics of human language that cyberneticians distanced themselves from many years ago. As I argued in the ASC column in the memorial issue for Francisco Varela (9:2), he made clear in his Calculus of self-reference that the triadic view of cognition as a self-organizing process is common in Peircean biosemiotics, second order cybernetics, autopoiesis, and enaction theory.

The theories of von Foerster and Maturana have had significant influence on the development of the Copenhagen school of biosemiotics. Conversely, Sebeok, who knew von Foerster, was very positive towards the idea of a cybersemiotics. In general he was supportive of enlarging the influence of semiotics, especially in fields where he could see that its knowledge could make a difference such as in artificial intelligence and cognitive science. Sebeok was a field maker in semiotics, and we are grateful for his help in generating the new field of cybersemiotics.

We are grateful to have some of the finest scholars on Sebeok's work and on biosemiotics, each of whom collaborated closely with Sebeok, contribute their thoughts to this issue. They are: Marcel Danesi, with whom he wrote his last book *The Forms of Meaning: Modelling Systems Theory and Semiotic Analysis* (2000, Mouton de Gruyter); John Deely, the semiotician and philosopher who wrote the tome *Four*

Ages of Understanding: The First Postmodern Survey of Philosophy from Ancient Times to the Turn of the Twenty-first Century (2001, Toronto University Press), where semiotics is welded into the history of philosophy in a complete new way building on the new perspective on semiotics that Sebeok developed; the biosemiotician Kalevi Kull who edited a volume on the modern significance of the work of Jacob von Uexküll for *Semiotica* [*134*(1/4), 2001]; and, finally, Susan Petrilli and Augusto Ponzio who co-authored of the shortest, most easily read, and yet scholarly books on Sebeok's work: *Thomas Sebeok and the Signs of Life* (2001, Icon and Totem Books).

Together the articles from these researchers give a very deep insight into the ideas behind the big version of semiotics based on Peirce instantiated as biosemiotics, a view you can also find argued in Deely's more demanding book *Basics of Semiotics* (Indiana University Press, 1990).

We have arranged the articles in an order that should be optimal for the reader with no or little prior knowledge of the semiotic paradigm. We also suggest that this issue might be used as course material for an introduction to the field of global- and biosemiotics that Sebeok created.

It is my hope that this special issue will promote the continuation of the work Sebeok began, and help inform scientists and scholars about the possibilities of the biosemiotic paradigm, especially the ways it can shed new light on problems within information science, cybernetics, system science and cognitive science. As there have already been many special issues and volumes on Sebeok's work within the "pure" semiotic community since he turned 80, I believe that creating this memorial issue in a transdisciplinary journal will be an inspiration to groups outside the semiotic community.

The obituary "Mister Biosemiotics" by Søren Brier dwells especially on Thomas Sebeok's contribution to the foundation of the field of biosemiotics. The ASC column: "The integration of second order cybernetics, autopoiesis and biosemiotics," is written by Søren Brier and discusses similarities and complementary aspects of cybernetics and biosemiotics. Marcel Danesi reviews Susan Petrilli's Italian biography of Thomas Sebeok.

The artist of the issue is Luciano Ponzio.

Modeling Systems Theory:
A Sebeokian Agenda for Semiotics

Marcel Danesi[1]

The traditional goal of semiotic theory has been to figure out how human semiosis unfolds, producing the multitude of signs that constitute the universe of meaning in cultural systems. The theoretical models of Ferdinand de Saussure and Charles Sanders Peirce stand, to this day, as frameworks for carrying out any type of discussion of human semiosis. Rarely has anyone ventured to reformulate their basic theoretical paradigm, or to link the study of human semiosis to semiosis across species. It was the late Thomas A. Sebeok who, in the last two decades of his life, started to formulate a new paradigm for semiotic theory, known as *biosemiotics*, that promises to reshape the semiotic research agenda in the future. Like the great biologist Jakob von Uexküll (1864-1944)—whose discovery by North American semioticians is due in large part to his efforts—Sebeok found a point of contact between the mainstream scientific approach to the study of organisms—*biology*—and that of the strictly *semiotic* cultural-philosophical tradition. Von Uexküll argued that every species had different inward and outward lives. The key to understanding this duality could be found in the anatomical structure of the species itself and in the kind of innate modeling systems it possessed. Animals with widely divergent anatomies do not have access to the same kinds of experiences and perceptions. In effect, a species does not perceive an object in itself, but according to its own particular kind of preexistent mental "modeling system" that allows it to understand it in a biologically-unique way. The essence of biosemiotics, as elaborated by Sebeok, is the perspective that semiosis is a concomitant of a species' modeling system. In this way, Sebeok has taken semiotics back to its biological roots, given that the discipline—at least in the Western world—grew out of the work of the Ancient Greek physicians. This paper will look at *modeling systems theory* as a serious proposal for broadening semiotic theory and method.

Introduction

For the sake of argument, the central goal of semiotics can be reduced, essentially, to figuring out the relation **[A stands for B]**, where **[A]** is generally called the *signifier* or the *representamen* and **[B]** the *signified* or *object*. The linkage of the two is known, of course, as the meaning-bearing structure called the *sign*. What constitutes a sign? And in what ways are the signs used by humans comparable to or different from the signs or signals used by animals?

While the first question has been the subject of much work and debate in semiotics, shaping its agenda from the time of St. Augustine to today, the latter question has rarely been entertained by mainstream semiotic theorists. It was the late Thomas A. Sebeok who argued persuasively that the two questions were intrinsically

---

1. Program in Semiotics and Communication Theory, Victoria College, University of Toronto, Toronto, Ontario M5S 1K7, Canada. E-mail: marcel.danesi@utoronto.ca. This paper is dedicated to the memory of Thomas A. Sebeok (1920-2001).

intertwined. And it was Sebeok who showed how the two could be broached in tandem, by proposing that the semiotic purview be expanded to include the study of signaling and communication systems in all species. It was his development of *Modeling Systems Theory* (MST), especially in several of his last works (e.g. Sebeok, 2001; Sebeok & Danesi, 2000), that has provided semioticians with a concrete framework for revisioning the **[A stands for B]** relation, and for conducting a study of semiosis across species. MST is based, basically, on the notion that the ability to manufacture *models* of the world through semiosis is a concomitant of biological life. Like the great biologist Jakob von Uexküll (1909), Sebeok argued that every species had different modeling systems, which were equipped to provide a specific species with a semiosic apparatus for understanding the particular type of biological world in which it existed and thus for coping with its particular form of existence (Sebeok, 1990, 1991, 2001).

The purpose of this memorial essay is to synthesize the main features of MST, since I believe that it can truly expand the scientific purview of semiotics. In a special 1999 issue of *Semiotica,* titled *Biosemiotica,* the basic Sebeokian paradigm was applied by a cadre of semioticians to their various fields of endeavor producing substantial theoretical fruits (see Nöth, 2001 for a review of the issue). If that tome is any indication, then the Sebeokian blueprint for semiotic research and theory development might indeed become the mainstream one in the near future. The reason for this, in my view, is that MST provides a notional framework that is more encompassing than the traditional one for carrying out investigations of semiosis.

Models and Forms

Semiotics emerged, arguably, as a distinct mode of inquiry in the study of physical *symptoms*. It was Hippocrates (c. 460-c. 377 BC) who identified the need to understand the bodily *semeion* as a physical sign, i.e., as an **[A stands for B]**, where **[A]** is a symptom and **[B]** the condition it represents or indicates. However, rarely in the history of semiotics has a truly biological approach to semiosis ever been envisaged, perhaps because rarely has a physician ever seen himself/herself consciously as a decoder of signs. It was Sebeok who argued throughout his career that the **[A stands for B]** relation should be studied in the same way that physicians study symptoms. Calling it *biosemiotics,* Sebeok wanted to refashion the primarily taxonomic Saussurean and Peircean traditions by expanding its purview to incorporate the study of semiosis across species

In Sebeok's biosemiotic framework, the notion of *modeling* is pivotal. This is definable as the species-specific ability to produce *forms* to stand for referents that have some relevance to species continuity. The use of the term *form* rather than *sign* or *signal* allows the theorist to integrate various semiosic phenomena that would otherwise have to be assigned to separate categories, even thought they may share properties. In the human species, forms may be purely imaginary, in which case they are equivalent to *mental images*, or they may be externalized—i.e. given material form—in which case they are *representations*. The work on sign theory in semiotics can in fact be reformulated as follows. When looked at globally, semiotics has occupied itself with studying four main types of signifying phenomena: (1) *signs* (words, gestures, etc.); (2) *texts* (stories, theories, etc.); (3) *codes* (language, music, etc.); and (4) *figural assemblages* (metaphors, metonyms, etc.). While there have been attempts to integrate these phenomena into one overall paradigm of semiosis, they have been, at best, scattered ones. Given the notion of *model* and *form* in MST, these can now be categorized as interrelated, given that they are all designed to model something in terms of the **[A stands for B]** relation. More specifically, a *model* can be defined as the overall relation **[A stands for B]** itself and a *form* as the **[A]** component of that relation, since it is something that has been imagined or made externally (through some physical medium) to stand for **[B]** (an object, event, feeling, etc.).

Model-making typifies all aspects of human intellectual and social life. Before building a house, a constructor will make a miniature model of it and/or sketch out its structural features with the technique of blueprinting. An explorer will draft a map of the territory he/she anticipates traversing. A scientist will draw a diagram of atoms and subatomic particles in order to get a "mental look" at their physical behavior. Miniature models, blueprints, maps, diagrams, and the like are so common that one hardly ever takes notice of their importance to human life; and even more rarely does one ever consider their *raison d'être* in the human species. Model-making constitutes a truly astonishing evolutionary attainment, without which it would be virtually impossible for modern humans to carry out their daily life routines. All this suggests the presence of a *modeling instinct* that is to human mental and social life what the physical instincts are to human biological life. Now, what is even more remarkable is that *modeling instincts* are observable in other species, as Sebeok has amply documented. The intriguing question that this reformulation of basic sign theory invariably raise is the following one: *What is the function of modeling in life forms?* This question begs, in turn, a whole series of related ones: *How is human modeling similar to, or different from, modeling systems in other species? What is the relation between modeling and knowing?* The purpose of biosemiotics is to seek answers to questions such as these by studying the manifestation of modeling behaviors in and across all species.

Modeling Systems Theory

For getting concrete data modeling systems within the human species and for comparing these across species, Sebeok has provided the following categories for assessing and thus evaluating the species-specificity of a given form (e.g., Sebeok, 1990):

- *Vocal forms:* This category refers to the fact that signals and messages can be transmitted vocally or nonvocally. Bird communication, for instance, is vocal; bee-dancing is nonvocal.

- *Verbal forms:* This category refers to the fact that verbal communication is unique to the human species. All other communication systems in Nature are nonverbal. Language is verbal, but not necessarily vocal (e.g. it can be communicated also by means of alphabet characters, gestures, etc.).
- *Witting forms:* This category refers to the fact that certain messages are unwitting or unconditioned (e.g. the signals sent out by pupil responses); others are witting, showing purposeful and intentional behavior.
- *Formation:* This refers to the fact that signaling, representational, and communication systems are *formed* in the organism by exposure to appropriate input in context and are subject to change or even *dissolution* over time. In all species, other than the human, forms are produced primarily through the biological channel; only human beings have the ability to create forms both through the biological channel and through cultural conditioning and exposure.

In MST, it is clearly necessary to distinguish between forms as they occur in Nature or in culture. A symptom is an example of an *externalized natural form*, i.e. a form produced by Nature. Words and symbols, on the other hand, are examples of *externalized artificial forms*, i.e. forms made intentionally by human beings to represent something. In the most recent work on MST (Sebeok & Danesi, 2000) four general types of artificial forms that humans are capable of producing have been identified: *singularized, composite, cohesive*, and *connective*. These serve many functions in human life. They allow people to recognize patterns in things; they act as predictive guides or plans for taking actions; they serve as exemplars of specific kinds of phenomena; and the list could go on and on.

In traditional semiotic theory *singularized forms* are called *signs*. In an MST framework, a sign can be defined, more precisely, as a form that has been made specifically to represent a singular (unitary) referent or referential domain. Singularized forms can be verbal or nonverbal. The English word *cat*, or the Spanish word *gato*, for example, are verbal singularized forms standing for the referent [carnivorous mammal with a tail, whiskers, and retractile claws]; a drawing of a house cat is its nonverbal (visual) equivalent. Now, a description of the same referent as *a popular household pet that is useful for killing mice and rats* constitutes, clearly, a different kind of form. This is known traditionally as a descriptive *text*. In MST, a text can be defined, more specifically, as a *composite form*; i.e. as a form that has been made to represent various aspects of a referent or referential domain—[household pet], [mice], etc.—in a composite (combinatory) manner. Classifying a *cat* in the same category as a *tiger, lion, jaguar, leopard, cheetah*, etc. exemplifies another type of modeling strategy—namely, the tendency to *codify* types of forms in some *cohesive* fashion. In MST, a code can be defined as a system that allows for the representation of referents perceived to share common traits—e.g. [cat], [tiger], [lion], [jaguar], etc. (= the feline code). Codes consist of interacting forms, forming a *cohesive* whole, which can be deployed to model types of phenomena in specific ways. Finally, the use of the word *cat* in an expression like "Alexander is a cool *cat*" is the result of a fourth type of modeling strategy, known traditionally as *metaphorical*. In MST, the term *connective form* is preferred instead, because a metaphor is a form that results, in effect, from the linkage of different types of referents (or referential domains): e.g. a human referent, [Alexander], with a feline referent, [cat]. These four types of modeling strategies are not mutually exclusive. Indeed, they are highly interdependent—signs go into the make-up of texts which, in turn, are dependent upon the elements that codes make available.

The ability to produce and understand models is innate. When an infant comes into contact with a new object, his or her instinctive reaction is to explore it with the *senses*, i.e. to handle it, taste it, smell it, listen to any sounds it makes, and visually observe its features. This exploratory phase of knowing the object constitutes a *sensory modeling* stage. The resulting internal model (mental image) allows the infant to *recognize* the same object subsequently without having, each time, to examine it over again "from scratch" with his/her sensory system (although the infant often will examine its physical qualities for various other reasons). Now, as the infant grows, he/she starts to engage more and more in semiosic behavior that replaces this sensory phase; i.e. he/she starts pointing to the object and/or imitating the sounds it makes, rather than just handling it, tasting it, etc. These imitations and indications are the child's first attempts at modeling the world in human terms (Morris, 1938, 1946). Thereafter, the child's repertoire of modeling activities increases dramatically, as he/she learns more and more how to refer to the world through the singularized, composite, cohesive, and connective modeling resources to which he/she is exposed in cultural context.

In effect, human development can be characterized as a process that starts with a *sensory modeling* phase that becomes more *conceptual* as the child grows. Concepts are, in effect, mental forms. As is well-known, two basic types of concepts have been distinguished traditionally in semiotics and philosophy—concrete and abstract. In MST, a *concrete concept* is defined as a form whose external referent is demonstrable and observable in a direct way, whereas an *abstract concept* is a form whose external referent cannot be demonstrated or observed directly. So, for example, the word *car* stands for a concrete concept because its referent, [a self-propelled land vehicle, powered by an internal-combustion engine], can easily be demonstrated or observed in the physical world. The word *love*, on the other hand, represents an abstract concept because, although [love] exists as an emotional phenomenon, it cannot be demonstrated or observed directly, i.e. the emotion itself cannot be indicated apart from the behaviors, states of mind, etc. that it produces.

In MST, it is assumed that the *form* that *knowledge* takes depends on the type of *modeling* used. As the psychologist C. K. Ogden and the literary critic I. A. Richards argued in their classic 1923 work, *The Meaning of Meaning*, it is sufficient to equate *meaning* with the particular *concept* elicited by a specific *representational form*. In traditional sign theory, the former is called the *signified*, and the latter, the *signifier*. Human representation, as Ogden and Richards observed, is a highly variable process. Like the indeterminacy involved in understanding natural phenomena, so too the exact nature of a *signified* is indeterminable in any objective sense, because its interpretation is shaped by situation, context, historical processes, and various other factors external to semiosis.

The types of forms discussed above are the end-results of three different, but interrelated, *modeling systems* present in the human brain, corresponding *grosso modo* to what Charles Peirce (1839-1914) called *firstness*, *secondness*, and *thirdness*. The child's earliest strategy for knowing an object with his or her senses is, in fact, a *firstness* strategy. The modeling system that underlies firstness forms of representation is the *primary modeling system* (PMS). The PMS can be defined as the instinctive ability to model the *sensory* or *perceptual* properties of referents. The child's subsequent attempts to refer to the object through vocal imitation and/or manual indication constitute a *secondness* knowing strategy. The modeling system that guides these attempts is the *secondary modeling system* (SMS). The SMS can be defined as the capacity to refer to objects with *extended* primary forms and with indexical (indicational) forms. Finally, in learning to use a culture-specific name to refer to an object, the child is engaging in a *thirdness* form of knowing. His/her ability to do so is dependent upon the *tertiary modeling system* (TMS), which can be defined as the capacity to acquire and utilize the symbolic resources of culture-specific abstract systems of representation.

Although MST has roots in the work of various twentieth-century structuralist semioticians, it has never really blossomed forth as a comprehensive theoretical and methodological framework for general use in theoretical semiotics until Sebeok's pivotal work. The elemental axiom, around which Sebeok has fashioned MST, is the conception that all semiosic phenomena can be categorized, as mentioned, in terms of four categories of forms—singularized, composite, cohesive, connective. From this axiom six principles follow:

1. Representation is the end-result of modeling systems producing forms of various types (the *modeling principle*).
2. Knowledge is indistinguishable from the forms used to encode it (the *representational principle*).
3. Modeling unfolds on three levels or dimensions, called primary, secondary, and tertiary (the *dimensionality principle*).
4. Complex (abstract) forms are derivatives of simpler (more concrete) ones (the *extensionality principle*).
5. Models and their referential domains are interconnected to each other (the *interconnectedness principle*).
6. All models and their forms display the same pattern of structural properties (the *structuralist principle*).

Needless to say, it is not possible to go here into the many interesting philosophical problems related to what is knowledge. The *representational principle* implies simply that in order for something to be known and remembered, it must be assigned some representational form. The *modeling principle* asserts that modeling is the activity that underlies representation. The *dimensionality principle* maintains that there are three dimensions or systems involved in modeling—primary, secondary, and tertiary. The *extensionality principle* posits that abstract forms are derivatives of more concrete, sense-based forms. The *interconnectedness principle* asserts that a specific form is interconnected to other forms (words to gestures, diagrams to metaphors, etc.). The *structuralist principle* claims that certain elemental structural properties characterize all modeling systems and forms. These are: *paradigmaticity, syntagmaticity, analogy, synchronicity, diachronicity,* and *signification*.

Paradigmaticity is a minimal differentiation property. To speakers of English, the two words *pin* and *bin* are kept distinct by a perceptible auditory difference in their initial sounds. This differentiation feature of sound systems is known in linguistics, of course, as *phonemic opposition*. Similarly, in classical Western music, a major chord is perceivable as distinct from a minor chord in the same key by virtue of a half tone difference in the middle tone of the chord. As such examples show, paradigmaticity is definable as the property of forms whereby some minimal feature is sufficient to keep them differentiated from all other forms of the same kind.

Syntagmaticity is a combinatory property of forms. Forms such as *tpin, tpill, tpit,* and *tpeak*, for instance, would not be legitimate words in English because the initial sequence [tp] + [vowel] is not characteristic of English word-formation, whereas words beginning with [sp] + [vowel] would: *spin, spill, spit, speak*. This combinatory feature of words is called, of course, *syllable structure*. Similarly, a major chord is recognizable as such only if the three tones are combined in a specific way: [tonic] + [median] + [dominant]. Syntagmaticity is definable as the property whereby the components of a form are combinable in some specifiable way.

Analogy is an equivalence property, by which one type of form can be replaced by another that is perceived as being comparable to it. The English word *cat* is analogous to the Spanish word *gato*; European playing cards can replace American cards if an analogy is made between European and American suits; Roman numerals can replace Arabic numerals through simple conversion; and so on.

Synchronicity refers to the fact that forms are constructed at a given point in time for some particular purpose or function; and *diachronicity* to the fact that they undergo change over time. The change is not random, but rather, governed by both structural tendencies characterizing the code to which it belongs and external contextual (social, situational, etc.) influences. As an example, consider the word *occhio* "eye" in Italian. The original form of this word was Latin *oculus*. Over time it became *oclu* (as various philological sources attest), and then *occhio*. These changes in physical form, however, did not come about haphazardly. The elimination of the middle vowel of *oculus (oclus)* and the subsequent change of *cl* to *cchi* (phonetically [kky]) were structural tendencies within the late (Vulgar) Latin phonemic system.

Finally, *signification* refers to the relation that is established between a form and its meaning. It is, more strictly, the relation that holds between the physical make-up of the form itself, the *signifier*, and the *referent* or *referential domain* to which it calls attention, namely the *signified*.

In a historical sense, the first true study of forms was the one carried out by St. Augustine (354-430 AD). This philosopher and religious thinker was among the first to distinguish clearly between *natural* and *conventional* (artificial) forms, and to espouse the view that there was an in-built *interpretive* component to the whole process of representation. But the task he laid out remained in virtual disregard until the writings of Saussure and Peirce. It is the work of the latter two which contained the foundational concepts for circumscribing an autonomous field of semiotic inquiry, aiming to study signs as elements related to each other systematically, rather than as isolated, material things in themselves. The key concept in semiotics has, in fact, always been that no single form can bear meaning unless it enters into systematic connections with other forms. The premise that biosemiotics adds to this basic paradigm is that the forms produced by a specific species are constrained by the *modeling system(s)* which has evolved from its anatomical constitution. The study of modeling systems thus involves a study in semiosis across life forms. For this reason, there are three main branches of biosemiotics, called *phytosemiotics*, the study of semiosis in flora (Krampen, 1981), *zoosemiotics*, the study of semiosis in fauna (e.g. Sebeok, 1963, 1968, 1972), and *anthroposemiotics*, the study of semiosis in humans.

It was, as well known, the British philosopher John Locke (1632–1704) who introduced the formal study of signs into philosophy in his *Essay Concerning Human Understanding* (1690), anticipating that it would allow philosophers to understand the interconnection between representation and knowledge. But the task he laid out, of discovering the properties of the sign, remained virtually unnoticed until Saussure and Peirce took it upon themselves to provide a scientific framework that made it possible to envision even more than what Locke had hoped for—namely, an autonomous field of inquiry centered on the sign (Deely, 2001). The subsequent development of semiotics as a distinct scientific domain, with its own methodology, theoretical apparatus, and corpus of findings, is due to the efforts of such twentieth-century scholars as Charles Morris, Roman Jakobson, Roland Barthes, A. J. Greimas, and Umberto Eco, to mention but a few.

The method of inquiry in biosemiotics is different from the traditional linguistic-philosophical approach basically in seeing commonalties among human and animal forms of semiosis. The study of anthroposemiosis requires special treatment because the most distinctive trait of human semiosis is that it permits both nonverbal—demonstrably derived from its primate ancestry—and uniquely verbal modeling. The forerunner of the biosemiotic movement was, as mentioned, the Estonian-born, German biologist Jakob von Uexküll (1864-1944) who was among the first to document manifestations of different types of semiosic behaviors exhibited by different phyla. The crux of von Uexküll's approach was his contention that every organism has different inward and outward modeling strategies for monitoring information.

Singularized Forms

The most basic type of human representational form is, of course, the sign. The sign is a singularized form because it constitutes a simple model serving to encompass a singular referent or referential domain. In anthroposemiosis, nonverbal signs include gestures, bodily postures, facial expressions, tones of voice, visual forms (e.g. drawn figures); verbal signs include words, intonation patterns, graphic signs (alphabetic, ideographic, etc.). In a biosemiotic paradigm, the function of singularized modeling is viewed as a general strategy for giving the perception of single objects, unitary events, individual feelings, etc. a knowable *form* (see also Thom, 1975; Sebeok, 1994). As von Uexküll (1909) argued, useful sensory information is something that a species would virtually not recognize, if it were not for the presence of an in-built modeling system designed to accomplish this task. Signs are, in effect, "recognition-enhancing forms," which allow for the detection of relevant incoming sensory information in a patterned fashion. Throughout the history of semiotics, there have been several attempts to identify and classify signs. Among these, Peirce's typology with 66 varieties is surely the most comprehensive, far-reaching, and sophisticated of all such attempts. In the verbal domain, one can also mention Roman Jakobson's (1970) classificatory system, which has shed considerable light on the minutiae of verbal modeling. Ignoring such minutiae for the sake of simplicity, six general categories of sign-making can be extrapolated from the relevant literature. These are: the *symptom*, the *signal*, the *icon*, the *index*, the *symbol*, and the *name*.

A *symptom* is a natural sign, recognizable by virtue of the fact that its form is coupled with its object inside the body's morphology. It is a manifestation of some altered physical (histological, cytological, etc.) process, ranging from a painful sensation (such as headache or backache), to a visible condition (such as a swelling or a rash), or change in body temperature. A group of symptoms that collectively characterize a disease or disorder is called a *syndrome*. A syndrome is, therefore, a composite form. It is a peculiarity of symptoms that their meanings are generally different for the patient (subjective symptoms) than they are for the physician (objective symptoms).

The *signal* is a singularized form that naturally or conventionally (artificially) triggers some reaction on the part of a receiver. Carpenter (1969, p. 44), a prominent researcher of animal behavior, defined signaling behavior as "a condensed stimulus event, a part of a longer whole, which may arouse extended actions." All animals are endowed with the capacity to use and respond to species-specific signals for survival. Birds, for instance, are born with the instinctive capacity to produce a particular type of coo, and no amount of exposure to the songs of other species, or the absence of their own, has any effect on their cooing behavior. A bird reared in isolation, in fact, will sing a very simple outline of the sort of coo that would develop naturally in that bird born in the wild. This does not mean, however, that animal signaling is not subject to environmental or adaptational factors. Many bird species have also developed regional cooing "dialects" by apparently imitating each other. A large portion of communication among humans also unfolds largely in the form of unwitting bodily signaling. Humans are capable as well of deploying witting signals for some psychosocial purpose—e.g. nodding, winking, glancing, looking, nudging, kicking, head tilting. As the linguist Karl Bühler (1934, p. 28) aptly observed, such signals act like social regulators, eliciting or inhibiting some action or reaction. Artificial, mechanical, or electronic signaling systems have also been created for conventional social purposes. The list of such systems is extensive, and includes: smoke signals, semaphores, telegraph signals, warning lights, flares, alarms, sirens, bleepers, buzzers, knocking, bells, etc.

A sign is said to be *iconic* when the modeling process employed in its creation involves some form of simulation. Iconic modeling produces primary singularized forms that display a perceptible resemblance between the form and its signified. In other words, an *icon* is a form that is made to resemble its referent in some way. Roman numerals such as I, II, and III are iconic forms because they imitate their referents in a visual way (one stroke = one unit, two strokes = two units, three strokes = three units); onomatopoeic words (*boom, zap, whack,* etc.) are also iconic forms because they constitute attempts to portray their referents in an acoustic way; commercially-produced perfumes that are suggestive of certain natural scents are likewise iconic because they attempt to simulate the scents in an artificial way; and the list could go on and on.

There are many manifestations of *iconicity* in zoosemiotic behavior as well, involving virtually all types of sensory channels—chemical, auditory, visual, etc. An elegant (if sometimes disputed) example of a complex form of signaling behavior that evolved, as it were, to function as a visual iconic form is graphically described by Kloft (1959). Kloft suggested that the hind end of an aphid's abdomen, and the kicking of its hind legs, constituted, for an ant worker, an iconic signifier, standing for the head of another ant together with its antennae movement. The ant can purportedly identify the likeness (the near end of the aphid) with its meaning (the front end of an ant), and act on this information, i.e. treat the aphid in the manner of an *effigy*, which is a visual icon.

A form is said to be *indexical* when its representational focus is the location of a referent in space, time, or in relation to some other referent. In one of his most memorable examples, Peirce referred to the footprint that Robinson Crusoe—the character created by British novelist Daniel Defoe (1660-1731) in his 1719 novel of the same name—found in the sand, which was interpreted by Crusoe as an index of some creature. In actual fact, a vast map of such indexical marks is printed overnight by animals of all sorts (Ennion & Tinbergen, 1967, p. 5). Indexes do not resemble their referents, like icons do; rather, they indicate or show where they are in relational terms. The most typical manifestation of *indexicality* is the pointing index finger, which humans the world over use instinctively to point out and locate things, people, and events in the world. Many words, too, manifest indexicality: e.g. *here, there, up, down,* etc.

A singularized form is *symbolic* when the modeling process employed in its creation is constrained by cultural and historical factors. Most semioticians agree that *symbolicity* is what sets human modeling apart from that of all other species, allowing human beings to represent things independently of stimulus-response situations. Many words are used symbolically. But any signifier—object, sound, figure, etc.—can be used symbolically: e.g. a cross figure can stand for the concept "Christianity;" a V-sign made with the index and middle fingers can stand for the concept "peace;" the color *white* can stand for "cleanliness," "purity," "innocence," and *dark* for "uncleanness," "impurity," "corruption;" and the list could go on and on. The ability to model the world symbolically is evidence that human consciousness is not only attentive to sensible properties (resulting in iconic modeling activities), and to spatiotemporal and relational patterns (resulting in indexical modeling activities), but also to all kinds of referents (actual and potential) in and of themselves.

A *name* is a form that identifies a human being (*Alexander, Sarah,* etc.) or, by connotative extension, an animal, an object (such as a commercial product), or event (such as a hurricane). A *name* has both indexical and symbolic properties: it is partly an indexical form because it identifies a person and, usually, points to his or her ethnic origin; it is partly a symbolic form because, like any word, it is a product of conventionalized representational practices. Less often, names are coined iconically. Trivial but instructive examples of this can be seen in the names given typically to household animals—*Ruff, Pooh-Pooh,* etc.

Composite Forms

Composite modeling is the activity of representing complex (non-unitary) referents by combining various forms in some specifiable way. Drawings, narratives, theories, conversations, etc. are all examples of composite forms. These are constructed with distinct signifiers that fit together structurally, but which are, as a whole, different from any of their constituent signifiers taken individually. In analogy to atomic theory, a singularized form can be compared to an atom and a composite form to a molecule made up of individual atoms, but constituting a physical form in its own right.

Texts are now definable as composite forms that incorporate the structural properties of the individual signifiers with which they are constructed, but they are not conceptually equivalent to the aggregate of their signifieds. A novel, for instance, is made up of words following one after the other. But conceptually it is not just the sum of the meanings of the words; rather, a novel constitutes a composite form that generates its own signified(s). Drawings, theories, and other composite forms are interpreted in this fashion. For instance, when asked what the theory of relativity is all about, people will typically couch their answer as follows: "The theory of relativity explains how time and space are interrelated." One can, of course, relate the signifying parts to each other in an interpretive discussion of the text. This is, in fact, what people do when they discuss a novel's meaning by referring to parts of the novel, to its plot, to its characters, etc. But in all such discussions, the parts are related to the signified extracted from it, rather than seen as separate from it.

There are as many types of composite forms as there are singularized ones. For example, syndromes are, as mentioned, composite forms which collectively indicate or characterize a disease, a psychological disorder, or some other abnormal condition. An example of an iconic composite form is an imitative drawing of a scene. An indexical composite form, such as a typical map, is one that is constructed to refer to spatial or temporal phenomena in an integrative relational way. A symbolic composite form, such as a mathematical theory, is a text that is made with the symbolic resources of a culture. Finally, a composite name consists of several identifiers (e.g. given name + surname) providing various kinds of culture-specific information—e.g. where the person is from, what his or her parentage is, etc. Composite modeling occurs in all facets of human life, allowing people to envision distinct bits of information and real-world phenomena as integrated wholes.

But composite modeling is not a specific capacity of human semiosis. It is, in fact, found in other species. One well-known example is the honeybee dance. Worker honeybees returning to the hive from foraging trips inform the other bees in the hive about the direction, distance, and quality of the food with amazing accuracy through movement sequences which biologists call a "dance," in obvious analogy to human dancing. The remarkable thing about the dance is that it appears to share with human representation the feature of displacement, i.e. of conveying information in the absence of the referential domain to which it calls attention.

Cohesive Forms

A cohesive modeling system is known in traditional semiotic theory as a code, a system providing particular types of signifiers that can be used in various ways and for diverse representational purposes. It can be compared to a computer program or to a common recipe. The former consists of a set of instructions that the computer can recognize and execute converting information from one form into another; the latter of a set of directions for preparing something to eat or drink by combining various ingredients. A language code, for instance, provides a set of phonetic, grammatical, and lexical "instructions" that the producers and interpreters of words and verbal texts can recognize and convert into messages. Generally speaking, for some particular representational need there is an optimum code or set of codes that can be deployed. For example, the composer of a work of operatic art will need to deploy at least three code-making sources in the construction of his or her text: the musical code, the verbal code, and the theater code (all in place at the time of the composition).

There are as many types of codes as there are signs or texts. For example, the body's immune system is a natural code consisting of interacting organs, tissues, cells, and cell products such as antibodies which not only neutralize potentially pathogenic organisms or substances, but also allow one to become aware of the difference between Self and "nonself" (the external world). It is the code that undergirds the symptomatology of diseases. An example of a simple mechanical (artificial) signaling code is the common traffic light system: a red light, green light, or yellow light inform a driver or pedestrian to stop, move forward, or slow down respectively. The Roman numeral system is an example of an artificial code fashioned in part iconically. This system consists of seven symbols for representing all numbers from 1 to 1,000,000: I for 1, V for 5, X for 10, L for 50, C for 100, D for 500, and M for 1000. The main iconic feature of this code is that one stroke represents one unit, two strokes, two units, three strokes three units: I = "one", II = "two," III = "three." An example of an indexical code is the system of street signs used typically to regulate and guide traffic. These signs provide information, among other things, about the distance of certain places from specific locations, about the direction one is traveling in, etc. An example of a simple symbolic code is the *Morse code*. This allowed people in the not-too-distant past to make verbal texts with dots and dashes (a dash is equal to three dots in duration) which were transmitted by a flash lamp, telegraph key, or other device. A letter or a number was represented (conventionally) by a combination of dashes and dots.

Connective Forms

Connective forms are the result of metaphorical reasoning processes. The ever-burgeoning literature on what has come to be known as *conceptual metaphor theory* (e.g., Lakoff & Johnson 1980, 1999; Lakoff, 1987; Johnson, 1987; Gibbs, 1994; Goatley, 1997) is highly intriguing, but still lacks a synthetic semiotic framework for interpreting the diverse, multiform manifestations of metaphor in human symbolic and communicative behavior. That framework is provided by MST and, more specifically, by the derived notion of *metaform*.

A *metaform* is an example of a *connective form* that results when abstract concepts are represented in terms of concrete ones. The formula [thinking = seeing], for example, is a metaform because it delivers the abstract concept of [thinking] in terms of the signifieds associated with the concrete concept of [seeing]. This metaform (which is a largely unconscious mental form in native speakers) underlies utterances such as:

7. I do not *see* what possible use your *ideas* might have.
8. I can't quite *visualize* what that new *idea* is all about.
9. Just *look at* her new *theory*; it is really something!
10. I *view* that *idea* differently from you.

Each of the two parts of the metaform is called a *domain*: [thinking] is called the *target domain* because it is the abstract topic itself (the "target" of the representation); and [seeing] is called the *source domain* because it enfolds the class of vehicles (forms with concrete signifieds) that deliver the meaning of the metaform (the "source" of the metaphorical concept) (Lakoff & Johnson, 1980). A specific metaphorical statement uttered in a discourse situation is now construable as a particular externalization of a metaform. So, when we hear people using such statements as those cited above, it is obvious that they are not manifestations of isolated, self-contained metaphorical creations, but rather, specific instantiations of the metaform whose target domain is [thinking] and whose source domain is identifiable as [seeing]:

The difference between a *metaform* and a *metaphor* is, in effect, one of hyponymy. A specific metaphor is a verbal instantiation of a metaform. Metaforms are *primary connective forms*, portraying abstractions in terms of concrete source domains. The [thinking = seeing] metaform, for instance, is linked to how we conceptualize [ideas], [theories], [awareness], [discernment], [clarification], [perspective], etc. These abstract notions are all conceived as *ways of seeing internally* that are modeled on *ways of seeing externally*.

Now, once the first "layer" of metaforms has been formed in a language's conceptual reservoir, on the basis of concrete source domains, then the layer becomes itself a new productive source domain for creating a higher (= more abstract) layer of concepts. This has been called the *layering principle* elsewhere (Danesi, 2001). The forms resulting from linkages among metaforms can be called *meta-metaforms*. Thus, for example, in utterances such as the following the target domain of [thinking] is rendered by source domains that are themselves metaforms: namely, [thinking = upward motion] and [thinking = scanning motion].

11. When did you *think up* that idea?
12. I *thought over* carefully your ideas.
13. You should *think over* the whole problem before attempting to solve it.

Phrasal verbs such as *think up* and *think over* are, in effect, products of an association of [thinking] with an [upward motion] and with a [scanning motion] respectively. A linkage of these two produces the meta-metaform: [thinking = upward + scanning motions] as in the following:

14. That idea *came out* of *nowhere*.
15. That theory *emerged* from the *landscape* of my thoughts.

Expressions such as *come out of nowhere* and *emerge from the landscape* are products of the meta-metaform [thinking = upward + scanning motions].

The third kind of connective form is really a type of symbol. For example, a *rose* is used as a symbol for *love* in Western culture because its physical features—rose = [sweet smell], [red color], [plant]—also constitute source domains for [love]: namely, [love = sweet smell], [love = red color], and [love = plant]. This is how the symbol [rose = love] came about. It is an example, therefore, of a *tertiary connective model*, which can be called, more specifically, a *meta-symbol*.

A metaform may also be the product of metonymic reasoning. *Metonymy* entails the use of an entity to refer to another that is related to it. A metonymic metaform results when part of a domain starts being used to represent the whole domain (Lakoff & Johnson 1980, pp. 35-40):

16. She likes to read Dostoyevski (= the writings of Dostoyevski).
17. He's in dance (= the dancing profession).
18. My mom frowns on blue jeans (= the wearing of blue jeans).
19. Only new wheels will satisfy him (= car).

Each one of these constitutes an externalization of a metonymically-derived metaforms (8) is an instantiation of [the author = his or her work], (9) of [an activity of a profession = the profession], (10) of [a clothing item = a lifestyle], and (11) of [a part of an object = the entire object].

Systems Analysis

Sebeok's framework for studying this nexus is called *Systems Analysis* (SA) (Sebeok & Danesi, 2000, p.123 passim). The main tasks of SA are: (1) to determine what constitutes a model in animal behavior, (2) to what modeling system it pertains (PMS, SMS, TMS), (3) what kind of modeling activity it manifests, and (4) what its function is. These tasks are guided by several key notions. First, there is the notion of dimensionality which, as mentioned, posits three distinct but interconnected types of models: (1) a *primary* model, which is a simulacrum of a referent; (2) a *secondary* model, which is either an extension of a simulacrum or an indexical form; and (3) a *tertiary* model, which is a symbolically-devised form of some kind. Second, there is the notion of stability vs. pliability which claims that a model (natural or artificial) can be *stable* (as in a written text) or *pliable* (as in oral conversation): stable models are fixed and relatively permanent or invariable; pliable ones are temporary and adaptive to the dynamics of a situation. Third, there is the notion which posits that the *form* a model assumes can be *singularized*, *composite*, *cohesive*, or *connective*, providing clues as to the nature of the referent or referential domain that it encodes. Fourth, there is the notion of interconnectedness, whereby the modeling system deployed will vary according to the nature of the referent, the function of the model, and the situation in which the modeling act occurs. Fifth, a distinction is made among semiosis, modeling, and representation: *semiosis* is the neurobiological capacity to produce forms (signs, texts, etc.), *modeling* is the channeling of the *semiosic* capacity towards a *representation* of some referent (the actual act of creating a form). Sixth, there is the notion that all models possess the same structural features (paradigmaticity, syntagmaticity, etc.). Finally, there is the notion that modeling reveals how the brain carries out its work of transforming sensory forms of knowing into internal forms of thinking and external forms of representation: a specific external *model* is thus considered to be a "cognitive trace" to the form a concept assumes in the mind, and since concepts depend on how they are modeled it has been argued throughout this book that the *form* that *knowledge* takes depends on the type of *modeling* used.

In SA, the species-specific forms of knowing are seen as manifest in the modeling behaviors of the species. Access to how a species knows something, therefore, is through the modeling system it possesses. Primary modeling, for instance, is "knowing through simulation." Secondary modeling, on the other hand, is "knowing through extension." This implies that the SMS does its handiwork, by and large, after the PMS has completed its own, in a manner of speaking. Further extensions of forms leads eventually to highly abstract, symbolic (tertiary) systems of representation (TMS). The PMS is the "default" system, while the SMS and TMS respectively are extensional systems.

SA would thus take systematically into its *modus operandi* the various facets of traditional semiotic analysis in an integrative fashion. Once the nature of the modeling process has been ascertained, then its forms and functions can be deduced or inferred from observation of the semiosic behavior involved. Thus, the cross-species nature of SA has clear implications for ethology and animal psychology, as well as for traditional semiotic theories and method.

Concluding Remarks

As Peirce (CP 1.538) cogently argued, "Every thought is a sign." But, as he also wrote, "Not only is thought in the organic world, but it develops there" (CP 5.551). This statement encapsulates why modeling is characteristic not only of the human world, but of the entire organic world, where, indeed, it developed. The *Umwelt* and *Innenwelt* of all animals, as well as the feedback links between the two, are created and sustained by the particular biology that characterizes a species. A model is a semiosic production with species-specific *formal* features.

As Sebeok showed throughout his illustrious career, the relatively simple, nonverbal models that animals produce are natural forms that must fit "reality" sufficiently to secure the survival and "sanity" of the members of a species in their ecological niche. In human beings, the modeling instinct is so pervasive and powerful that it often becomes very sophisticated indeed in the adult life of some individuals, as borne out by Einstein's testimonial, or by what we know about Mozart's or Picasso's ability to model intricate auditory or visual referents in their heads in anticipation of transcribing them onto paper or canvas. Language, metaforms, meta-symbols, as far as we know, are unique to anthroposemiosis. These make it possible for humans not only to represent immediate reality, but also to frame an indefinite number of possible worlds. The modeling capacity in humans has led to what Bonner (1980, p. 186) calls "true culture," requiring "a system of representing all the subtleties of language," in contrast to "nonhuman culture." It is on this level, defined as tertiary, that nonverbal and verbal sign assemblages blend together in the most creative modeling system that Nature has thus far produced.

There are many questions that MST raises for semiotics. But, I would like to conclude on a personal note, by claiming that I believe that it is probably headed in the right direction and that the questions can be tackled along the way. I had the pleasure of collaborating with Sebeok in drafting a biosemiotic agenda for semiotics (Sebeok & Danesi, 2000) just before he passed away. I was at first doubtful that it could offer something significantly different from the traditional Saussurean-Peircean paradigm. I am now convinced that Sebeok was on the right track. In having attempted to rebuild semiotics as a "life science," taking it back, in effect, to its roots in biology, he left us an exciting new agenda for conducting research that is truly interdisciplinary and apt to produce interesting and meaningful results. The attractive aspect of MST, in my view, is that it allows us to use a standard terminology for studying semiosis across species, which in turn, allows us to establish a taxonomy of notions, principles, and procedures for understanding the uniqueness of human semiosis. The result will be, in my estimation, a rigorous program for studying human cognition as a capacity that transforms sensory-based and affectively-motivated responses into a world of mental models, and then to reconstruct that world through them whenever we want. As Kull (2001, p. 94) has recently put it, the biosemiotic movement will produce outcomes that produce a "better understanding of life itself."

References

Bonner, J. T. (1980). *The evolution of culture in animals*. Princeton: Princeton University Press.
Bühler, K. (1934). *Sprachtheorie: Die Darstellungsfunktion der Sprache*. Jena: Fischer.
Carpenter, C. R. (1969). Approaches to studies of the naturalistic communicative behavior in nonhuman primates, In T. A. Sebeok et al. (Eds.), *Approaches to animal communication* (pp. 40-70). The Hague: Mouton.
Danesi, M. (2001). Layering theory and human abstract thinking. *Cybernetics & Human Knowing, 8*, 5-24.
Deely, J. (1989). A global enterprise: Preface to *The sign and its masters* by Thomas A. Sebeok, pp. vii-xiv. Lanham: University Press of America.
Ennion, E. R., Tinbergen, N. (1967). *Tracks*. Oxford: Oxford University Press.
Gibbs, R. W. (1994). *The poetics of mind: Figurative thought, language, and understanding*. Cambridge: Cambridge University Press.
Goatley, A. (1997). *The language of metaphors*. London: Routledge.
Jakobson, R. (1970). Language in relation to other communication systems. In C. Olivetti (Ed.), *Linguaggi nella società e nella tecnica* (pp. 3-16). Milano: Edizioni di Comunità.
Johnson, M. (1987). *The body in the mind: The bodily basis of meaning, imagination and reason*. Chicago: University of Chicago Press.
Kloft, W. (1959). Versuch einer Analyse der Trophobiotischen Beziehungen von Ameisen zu Aphiden, *Biologische Zentralblatt, 78*, 863-870.
Krampen, M. (1981). Phytosemiotics. *Semiotica, 36*, 187-209.
Kull, K. (2001). Living forms are communicative structures, based on the organic codes. *Cybernetics and Human Knowing, 8*(3), 91-94.
Lakoff, G. (1987). *Women, fire and dangerous things: What categories reveal about the mind*. Chicago: University of Chicago Press.
Lakoff, G., & Johnson, L. (1980). *Metaphors We Live By*. Chicago: Chicago University Press.
Lakoff, G., & Johnson, M. (1999). *Philosophy in the Flesh: The Embodied Mind and Its Challenge to Western Thought*. New York: Basic.
Nöth, W. (2001). Biosemiotica. *Cybernetics and Human Knowing* 8/1-2: 157-160.
Ogden, C. K., & Richards, I. A. (1923). *The Meaning of Meaning*. New York: Harcourt, Brace.
Peirce, C. S. (1931-1958). *Collected papers*. Cambridge, MA.: Harvard University Press.
Saussure, F. de. (1916). *Cours de linguistique générale*. Paris: Payout.
Sebeok, T. A. (1963). Communication in animals and men. *Language, 39*, 448-466.
Sebeok, T. A. (1968). *Animal communication: Techniques of study and results of research*. Bloomington: Indiana University Press.
Sebeok, T. A. (1972). *Perspectives in zoosemiotics*. The Hague: Mouton.
Sebeok, T. A. (1976). *Contributions to the doctrine of signs*. Lanham: University of America Press.
Sebeok, T. A. (1979). *The sign and its masters*. Austin: University of Texas Press.
Sebeok, T. A. (1981). *The play of musement*. Bloomington: Indiana University Press.
Sebeok, T. A. (1985). Pandora's box: How and why to communicate 10,000 years into the future. In M. Blonsky (ed.), *On signs*, (pp. 448-466). Baltimore: Johns Hopkins University Press.

Sebeok, T. A. (1986). *I think I am a verb: More contributions to the doctrine of signs*. New York: Plenum.
Sebeok, T. A. (1990). *Essays in zoosemiotics*. Toronto: Toronto Semiotic Circle Monograph Series.
Sebeok, T. A. (1991). *A sign is just a sign*. Bloomington: Indiana University Press
Sebeok, T. A. (1994). *Signs: An introduction to semiotics*. Toronto: University of Toronto Press.
Sebeok, T. A. (2001). *Global semiotics*. Bloomington: Indiana University Press.
Sebeok, T. A. and Danesi, M. (2000). *The forms of meaning: Modeling systems theory and semiotics*. Berlin: Mouton de Gruyter.
Thom, R. (1975). *Structural stability and morphogenesis: An outline of a general theory of models*. Reading: W.A. Benjamin.
Uexküll, J. von (1909). *Umwelt und Innenwelt der Tierre*. Berlin: Springer.

The Quasi-Error of the External World
an essay for Thomas A. Sebeok, in memoriam

John Deely[1]

Betwixt and Between

There is a story according to which Professor Sebeok was on a panel of distinguished speakers who received from the audience a challenge to show cause why the basic ideas of semiotics, such as that of Umwelt, were not simply one more version of solipsistic idealism. Each of the speakers in turn addressed the matter, each beginning with a protestation (outdoing in earnestness the previous speaker) to the effect that, "Of course, I am not a solipsist." Finally, Tom's turn arrived. He shrugged, and said simply: "I'm a solipsist." It was one of those seminal moments, of which Tom created so many, like the time in Toronto where he mentioned in passing in his main remarks that "Everyone thinks of language in terms of communication. But language has nothing to do with communication." In the question period, the very first questioner challenged him on the point. "You said that language has nothing to do with communication," the audience member reminded him. "Why did you say that?" "Because it doesn't," Tom answered pointedly,[2] and proceeded to call on the next questioner.

The Egg of Postmodernity

It was fascinating, to borrow Tom's own description of a former instance of the type of event in question,[3] "to note a clue *in nuce*, lurking" in these calculated rhetorical outrages; but even more fascinating was it to watch these clues unfold into full-blown theoretical insights over the following months and years, floating like water lilies on seas of detail.

Now Tom Sebeok was a man of details. Even though he never assumed for himself the mantle of philosopher, as the twenty-first century advances, it will prove difficult, if not impossible, for professors and students of philosophy to maintain ignorance of his name. Mayhap no single man of the twentieth century, save perhaps

---

1. University of St. Thomas, Houston, Texas. Center for Thamistic Studies, 3800 Montrose Blvd. Houston, TX 77006. E-mail: deelyj@stthom.edu
2. Later Sebeok would say (1995, p. 91, or 2001, p. 70), in the form of a summary maxim, as from a medicine man to the tribal elders, "Resist the temptation to jumble three incommensurate semiosic practices and their corresponding appellations: communication, language, and speech."
3. Sebeok, 2000, p.143. He was referring to the 1969 oral presentation of Langer 1971, in which he saw a glimmer of his later distinction between language as component of the species-specifically human modeling system and its exaptation to create linguistic communication. It is beginning to be "in the air," perhaps: cf. Ashley 1985, p. 40.

Peirce himself (though in quite a different and less collegial fashion!), had as shaping an influence on the intellectual culture and climate of what is destined, in my opinion, to be called with a positive sense the "postmodern era."

For "postmodern" has a very different meaning as it bears on philosophy than has heretofore generally been suspected. For example, Karol Wojtyla noted that the term "postmodernity," as it began to find currency in the late twentieth century, was first used with reference to aesthetic, social and technological phenomena. The term was then transposed into the philosophical field, "but has remained somewhat ambiguous," mainly because "there is as yet no consensus on the delicate question of the demarcation of the different historical periods."[4] Interesting pretensions to the contrary notwithstanding,[5] in the end, the consequence that "postmodern" supersedes "modern" is unavoidable.

Well, "the delicate question of the demarcation of the different historical periods," I make bold to say, has been recently addressed at length. Modernity, in philosophy, means that period, beginning with the *Meditations* of Descartes, which came to assume in its mainstream development that the products of the mind's own workings provide alone the direct objects of experience on the side of consciousness (See Deely, 2001, Chapters 11–14). Locke shared this supposition with Descartes, and Kant did not challenge it, even though he introduced into the notion of consciousness a structure of relationality which considerably distanced him from his modern forebears, but nonetheless without overturning the essential tenet of idealism: what the mind knows the mind makes.

Now there is hardly room for doubt at this juncture of history about a social construction of reality, that human consciousness structures human experience of objects, and perhaps even that the main contours of what we experience is a product more of our thinking and sociocultural conventions than of any input from a "nature" recognized as such that is operative within our experience as well as prior to, and independently of, that experience. "Reality" for the traveler today, is less a matter of undiscovered lands and seas than a matter of correct papers of identification and negotiation of officials at travel and custom points along the various frontiers. These boundaries, moreover, themselves depend upon diverse traditions, so that the boundaries of nations today are not those of a thousand years ago, nor of a thousand years hence. Even the Pope of Roman Catholicism, an icon of a spiritual reality

---

4. Wojtyla, 1998, ¶ 91: "A quibusdam subtilioribus auctoribus aetas nostra uti tempus «post-modernum» est designata. Vocabulum istud, saepius quidem adhibitum de rebus inter se dissidentibus, indicat emergentem quandam elementorum novorum summam quae sua amplitudine et efficacitate graves manentesque perficere potuerunt mutationes. Ita verbum idem primum omnium adhibitum est de notionibus ordinis aesthetici et socialis et technologici. In provinciam deinde philosophiae est translatum, at certa semper ambiguitate signatum, tum quia iudicium de iis quae uti «post-moderna» appellantur nunc affirmans nunc negans esse potest, tum quia nulla est consensio in perdifficili quaestione de variarum aetatum historicarum terminis. Verumtamen unum illud extra omnem dubitationem invenitur: rationes et cogitationes quae ad spatium post-modernum referuntur congruam merentur ponderationem."
5. Lyotard, 1984, p.79: "A work can become modern only if it is first postmodern. Postmodernism thus understood is not modernism at its end but in the nascent state, and this state is constant." How nice to realize that modernity, after all, is an eternal condition.

transcending humanity, for many centuries now, is chosen by election among "Cardinals." Yet no one doubts that cardinals are wholly a creation of human tradition within that Church, no different as objective realities than the Electors in the Electoral College of the United States, or the former Electors of the erstwhile "Holy Roman Empire."

So the objective world of human experience is at best a mixture of nature and culture, but a mixture in which the predominant formal patterns come more from culture than from nature. These formal patterns from culture underlie the presentation to us of objects directly experienced. The situation in this regard in fact was no different for the ancients or medievals; but they had not awakened to the fact. Who are our relatives? It depends on what kinship system is regnant in the culture in which we are raised. What is our religion? Allowing for individual exceptions, again the answer depends mainly on the historical circumstances into which we are born and raised.

The moderns, awakening to all this, had good reason to see in the objects experienced creations of the mind's own workings. Even science might be reduced to the same: such was the Kantian experiment, undertaken from the impetus of Hume's discourses maintaining nothing more to experience than customary associations among objects. If all thoughts reflect mere habits, and all objects cannot be known to be anything other or more than mental self-representations, whatever suspicions we may have on "common sense" grounds that there is a world independent of us, skepticism is the final warrant of all knowledge. That, to Kant, was unacceptable. He could swallow everything else his modern mainstream forebears taught him, but not skepticism. Their mistake, as he saw it, was in reducing knowledge to subjectivity — that is to say, their mistake was in identifying ideas in the individual mind with the objects of direct experience. They have failed to grasp that knowledge is essentially relational in structure, and that relations are over and above the subject. So, ideas in the individual give rise to or "found" cognitive relations to *object*s. But the objects are that at which the relations terminate, not that upon which the relations are founded and whence they provenate. And the manner in which these relations are generated in giving form to objects is according to a pattern in-built, a-priori to, the human mind, not a mere matter of habits of association.

Now, when it comes to the objects of specifically scientific knowledge, according to Kant, we are dealing with a universality and necessity which comes from the mind itself, not some mere habit pattern and customary generalization. Even though the external world remains unknowable in its proper being, still we know *that* it is there; and our manner of thinking it intellectually is not capricious or culturally relative but universal and necessary, the same for all humans. Even though we know only what our representations give us to know, and the representations are wholly of our mind's own making, still they are made into objects not by association but a-priori, independently of the vagaries of custom and individual experience; and their cognized content is not subjective but objective, that is to say, given at the terminus of relations which our representations only found. The scientific core of human experience, contrary to the animating conviction of scientists themselves, mind you,[6] but according to the modern philosopher Kant, is prospectively the same for all because the sensory mechanism generating the representations involved in it and the conceptual mechanism organizing relations arising from these representations is the same in all: similar causes result always in similar effects. Such was Kant's version of the medieval adage, *agens facit simile sibi*.

By this simple expedient Kant thought to have settled the objectivity of knowledge and put skepticism at bay. The scandal of not being able to prove that there is a world external to the human mind had been removed by the proof that indeed there is an unknowable realm to which knowledge cannot extend, and this realm is precisely the external world, known of a certainty to exist as stimulating our representations (in sense intuition) and unknowable in itself (through concepts, which yield knowledge only in correlation with the representations of sensory intuitions but result in yet another realm of unknowables, the noumena, if we try to extend them beyond the boundary of what is represented in intuitions of sense).

Later variants on this Kantian theme would concentrate on the phenomena in our experience of objects, after Husserl. Yet others would concentrate on the language itself in which human representations are mainly systematized, after Frege, Russell, and Wittgenstein. Russell, indeed, emphasized the role of relations in language and (hence) in knowledge. But Kant had already done this(See the analysis in Deely, 2001, pp. 561–63); and Russell confessed in the end that he, too, was unable to transcend solipsism:

> My own belief is that the distinction between what is mental and what is physical does not lie in any intrinsic character of either, but in the way in which we acquire knowledge of them. ... I should regard all events as physical, but I should regard as *only* physical those which no one knows except by inference. (Russell, 1959, p. 254)

We are linked to other minds in exactly the same manner by which we are linked to any reality external to ourselves: by inference. Directly we experience only what is in ourselves. Such was modernity when Sebeok came along. Such was modernity when he left it.

The Egg Hatches

This conundrum, this Gordian knot of "external reality," with which modernity had paralyzed the philosophers, was precisely what Sebeok cut. Faced with the dilemma of the modern implication of a maze of the mind's own workings, he found the way out:

> What is one to make of all this? It seems to me that, at the very least for us, workers in a zoösemiotic context, there is only one way to get through this thicket, and that is to adhere strictly to Jakob von Uexküll's comprehensive theses about signs. (Sebeok, 2001, pp. 78 & 193n6)

---

6. We will have further occasion to expand on this point. Cf. Sokal & Bricmont, 1998.

Of course, Sebeok is speaking here about *Umweltstheorie*, nothing less, to which nothing is more central than Sebeok's own distinction between language in the root sense (that aspect of the human modeling system or *Innenwelt* that is underdetermined biologically and as such species-specific to us) and the exaptation of language to communicate (resulting in linguistic communication as, again, a species-specifically human *moyen*). The mistake of the cultural relativists of whatever stripe, and of the moderns in general, from Sebeok's point of view, was to treat of language as an autonomous system all-encompassing, instead of to realize its zoösemiotic context (Deely, 1980). In this context "the uniqueness of man" stands out only insofar as it is sustained by and depends upon commonalities of signifying that define and constitute the larger realm of living things within which human beings are perforce incorporated and bound up by a thousand million lines of relationships which the very understanding of human life must ultimately bring to some conscious incorporation. In other words, Sebeok's final contribution, from within modernity, was to realize that there was a way beyond modernity, the Way of Signs.

The question in this form never interested him, but in the wake of his work it is worth asking: what would postmodernity have to be? It could only be a view of the world which somehow managed to restore what is external to ourselves as knowable in its own or proper being without letting go of or denying the modern realization that much if not most of what we directly know reduces to our own customs and conventions according to which objects are structured and inferences made. Now Sebeok may indeed have been thoroughly modern. But no one did more than he to ensure that the modern era, at least in intellectual culture and in the philosophy of which he never claimed to assume the mantle, was over. Sebeok became, in spite of himself, postmodern to the core.

I say "in spite of himself," for I know he had an aversion to the designation "postmodern," for a very good reason. The greater part of Sebeok's professional life had been devoted to exposing and overcoming what has come to be known as the "*pars pro toto* fallacy,"[7] according to which the doctrine of sign finds its adequate foundation on a linguistic, not to say verbal, paradigm. Such was the thesis, well known by the twentieth century's end, of Saussure's proposal for "semiology." By contrast, Sebeok promoted from the first that the doctrine of signs must be rooted directly in a general study of the action of signs, the distinctive manner in which signs work, for which action he accepted from Peirce the name "semiosis." But if the doctrine of signs concerns first of all the action revelative of the distinctive being of signs, then the proper proposal for its development is not the term "semiology" but rather the term "semiotics," which expresses the ideal of a paradigm not language-bound and refers to the action of signs as larger than, surrounding, and indeed presupposed to any action of signs as verbal. "Semiosis," Sebeok said early on, "is a pervasive fact of nature as well as of culture" (1977a, p. 183). Semiology he always

---

7. "Pars Pro Toto," Preface to Deely, Williams, & Kruse, 1986, pp. viii–xvii; also in Deely, 1986c.

saw as having a legitimate place within semiotics — the glottocentric part of the larger enterprise, as he put it, but impossible to be the whole.[8]

You can see that Sebeok from the first, without at all thinking of the matter in these terms, set the semiotic enterprise beyond the philosophical boundaries of modernity. His was a postmodern enterprise, from the philosophical point of view, right from the start, however gradually has this fact come into the light. Two things conspired to make it difficult for him (as for us around him) to recognize this. The first was that, by his own profession (what turned out to be to his great advantage), he had not made his forté philosophy. The boundary of modernity as a philosophical epoch had not yet been clearly drawn in the days when Sebeok set out along the Way of Signs, even though Sebeok's own works within linguistics, anthropology, and folklore, as well as later in semiotics, were pushing intellectual culture in the direction of a becoming conscious of just that boundary, as well as of a way across it. The second, much more immediate reason was that, within the larger orbit of semiotics, a group of thinkers vaguely semiological, certainly glottocentric, with Jacques Derrida at their center,[9] had clustered in the consciousness of popular culture around the label "postmodern." The general intellectual thrust of this group, besides being "abjectly based on the *pars pro toto* fallacy," was, to Sebeok's intellectual sensibilities (not to put too fine a face on it), abhorrent. Viewing this development within semiotics and popular culture more generally, Sebeok once confided to me that he deemed the appellation *postmodern* as "so hopeless as better never to be used."[10]

Yet the term has a logic of its own. Since that which every object as experienced ("real" as the sun or "unreal" as the witches of Salem) presupposes is the action of signs, the way out of the closet of modernity's solipsism is not by going back to a simple ancient or medieval realism but rather by going forward into a brave new world. *Faute de mieux* the new world will be, and not only for philosophy but for intellectual cultural in its trajectory over-all, *postmodern*. Sebeok's objection to those already identified with the label *postmodern* in end-of-the-twentieth-century popular intellectual culture was their abject manipulation of the glottocentric model, their whole-sale debasement of the possibilities proper to semiology as a normal part within the larger doctrine of signs. It did not occur naturally to Sebeok that this situation was not by any means postmodern in any philosophical sense. In a philosophical sense, the situation in question was rather *ultramodern*, the simple carrying to the extreme of the modern proposition that the mind knows only what the mind makes. The "postmoderns falsely so called" were not postmodern at all: they were philosophical

---

8. E.g., see the "Introduction" to Sebeok, 2001, especially pp. xix–xxiii. It was always an irritant to Sebeok that Greimas, a glottocentrician if ever there was one, claimed for his portentously-named "Paris School" the designation "semiotics" rather than the far more apt title of a school of "semiology."
9. Outside the orbit of Continental semiology, the late-modern "pragmatism" of Richard Rorty performed on the American scene the same function of extending the modern twilight even into the postmodern dawn.
10. Conversation circa1984. He was not so far removed on this point from the devastating remarks of Sokal & Bricmont (1998) who, however, made the same mistake: to wit, to accept at face-value ("nominalistically," as it were) the claim that thinkers styled "postmodern" have ipso facto just claim to the label.

modernity extended to the extreme that Kant had essayed to forestall: swamping science itself in a linguistic tangle of terminally clever solipsistic relativism.

If modern philosophy depends upon an epistemological paradigm which knows no path beyond the representative contents of consciousness, while epistemology constitutes for semiotics no more than its "midmost target" (Sebeok, 1991, p. 2), the reason is that study of the action of signs finds precisely a path[11] beyond the representative contents of consciousness. These contents are not self-representations (objects) but, precisely, themselves signs (other-representations)[12] rooted in the being of relations which transcend the division between nature and culture, inner and outer, and so cannot be confined to either side of any such divide, real or imagined. The "central preoccupation" of semiotics may be, à la modernity, "an illimitable array of concordant illusions," Sebeok reported to the Semiotic Society of America in his Presidential Address of 1984, but "its main mission" is "to mediate between reality and illusion" (Sebeok, 1984, pp. 77-78). Petrilli and Ponzio, in their recent study of Sebeok's work (which had something of his endorsement), capture the postmodern essence of the way of signs as Sebeok envisioned it exactly: "there is no doubt that the inner human world, with great effort and serious study, may reach an understanding of non-human worlds and of its connection with them" (Petrilli & Ponzio, 2001, p. 20).

Skirmishes on the Boundary

When the postmoderns falsely so called entered the fray from the fringes of semiology, just when semiotics was promising to come into its own, Sebeok could not but be dismayed. For Sebeok, insofar as he was modern, belonged to the scientific, not the philosophical, side of modernity. So it needs also to be noted, as mentioned above in passing, that modern intellectual culture became frankly and unreservedly idealistic only on its philosophical side, and this even in spite of itself. Neither Galileo nor Descartes set out to make the external world problematic, still less unknowable. These were rather the consequences ineluctably radiating from the modern starting point which only the philosophers wholeheartedly embraced, to wit, the premiss that we directly know nothing but self-representations fashioned by the mind under the provocation of stimuli directly unknown. A realistic spirit never wholly died, neither within the popular culture as a residual "common sense," nor within the scientific enterprise with which modernity thought to replace coenoscopic knowledge with a wholly ideoscopic edifice (the "Enlightenment" project), even if only to learn ruefully that, after all, coenoscopy has its irreducible place in the realm of knowledge alongside and in some ways naturally prior to ideoscopy.[13] Modern science took its

---

11. "Renvoi," as has been said (after Jakobson, 1974), Deely, 1993a, together with the further revision proposed in Deely, 2001e, pp. 721–22.
12. The point is fundamental: cf. Poinsot, 1632, p. 117/12–17. Cf. the discussion of the status of the representamen in Deely, 2003.
13. See Deely, 2001, Chapter 11, esp. pp. 489–492; cf. Heidegger, 1927, pp. 10–11, on how "ontological inquiry," provided it does not remain "naïve and opaque" in its researches after the manner of Galileo's critics and judges at the time, "is indeed more primordial, as over against the ontical inquiry of the sciences."

origin not from a repudiation of but in continuity with the ancient and medieval concern with exposing in knowledge the very structures and modalities proper to *ens reale*, the order of things-in-themselves.[14] Galileo's social problems arose not from proposing hypotheses about what might be but from proposing a hypothesis about *the way things are*, in the spirit and with the conviction that such was knowable, even if it required new instruments and different means than were developed by or available to the medieval "natural philosophers" and early modern (or any other) religious authorities.

So the modernity upon the scene of which Sebeok entered and of which he was one of the most noble heirs of its high intellectual culture had a schizophrenic, not to say psychotic, side, for the purpose of understanding which I have proposed we might usefully exapt the story of Dr. Jekyll and Mr. Hyde (Deely, 2001, Chapt. 13). Fortunately for all of us, Sebeok was a man of thoroughly scientific temper, and while he had respect for philosophy and its conundrums, and even while the sciences in which he immersed himself concerned objects which could be in nowise reduced to the *ens reale* at the heart of the original enterprises of philosophy and science alike, neither was he about to be taken in by a central thesis that served to define little more than philosophy's distinctively modern mainstream. He was never one for abjectly mistaking some part for the whole, or for respecting the claims of those who transparently did so, such as the ultramoderns, "the postmoderns falsely so called."

"For us, workers in a zoösemiotic context, there is only one way to get through this thicket"(Sebeok, 2001, pp. 78 & 193n6). Realism is not enough. The Way of Things has been tried and found wanting, even if not so completely so as the Way of Ideas proved wanting. For along the Way of Signs, we find that realism, "scholastic realism," as Peirce insisted ("or a close approximation to that"[15]), if insufficient, yet pertains to the essence of the enterprise. The only way to realism in the minimalist sense required (the sense of "scholastic realism," that is, as Peirce correctly termed it) is by a warranting within experience of a distinction not only between signs and objects (the former as presupposed to the latter's possibility), but further between objects and things. Here "things" refers to a dimension within the experience of the objective world which not merely does not reduce to our experience of it but is further knowable as such within objectivity through the discrimination in particular cases of what marks the difference between *ens reale* and *ens rationis* in the being of objects experienced. In this way (and for this reason), the "minimal but sufficient module of distinctive features of +, −, or 0, variously multiplied in advanced zoösemiotic systems" which yet remain wholly perceptual in nature, "is a far cry from the exceedingly complex cosmic models Newton or Einstein in due course bestowed upon humanity"(Sebeok, 1995, p. 87, or 2001, p. 68).

---

14. I would direct the reader's attention to remarks I made on this subject at one of the many conferences organized by Sebeok: (Deely, 1984, pp. 265–66).
15. Peirce, 1905, CP 5.423. By *scholastic realism* Peirce intends in general a sense of realism sufficiently strong and clear as to prove incompatible with all variants of Nominalism as the denial of relations sometimes obtaining in their proper being as relations independently of the workings of finite mind.

And the only way to a difference between things experienced as objects through their relation to us and things understood, actually or prospectively, as things in themselves prior to or independent of any such cognitive relation is through a modeling system capable of proposing within some object of experience a difference between aspects of the object given in experience and those same aspects giveable apart from the particular experience. In Sebeok's trenchant terms, the distinction between objects and things depends upon a modeling system, an *Innenwelt*, which has among its biologically determined components a component which is biologically underdetermined, the component Sebeok labels "language." Thanks to language we can model the difference between "appearances" in the objective sense and "reality" in the scholastic sense, and propose experiments to test the model, leading to its extension, refinement, or abandonment, depending upon the particulars of the case.

An Umwelt species-specifically human is not needed in order for the mind to be in contact with reality in this scholastic sense.[16] For reality in the scholastic sense need not be envisioned in order to be objectively encountered. The *idea* of reality, we will shortly see, is no less than a representative component of an Innenwelt species-specifically human. But the *objective content* of that idea (not, indeed, formally prescissed as such, but, indeed, "materially" in the scholastic sense) is part of the Umwelt of every animal.[17] Sebeok liked to quote Jacob on the point:

> No matter how an organism investigates its environment, the perception it gets must necessarily reflect so-called 'reality' and, more specifically, those aspects of reality which are directly related to its own behavior. If the image that a bird gets of the insects it needs to feed its progeny does not reflect at least some aspects of reality, then there are no more progeny. If the representation that a monkey builds of the branch it wants to leap to has nothing to do with reality, then there is no more monkey. And if this point did not apply to ourselves, we would not be here to discuss this point. (Jacob, 1982, p. 56)

REALITY TOO IS A WORD

So a *grasp* of reality is not the issue. The issue is much deeper, and surely begins with the realization concerning which Sebeok was fond of citing Niels Bohr: "'Reality' is also a word, a word which we must learn to use correctly"(French & Kennedy, 1985, p. 302). Now, as a word, *reality* has a history, one which, in philosophy at least, leads

---

16. What Peirce calls "scholastic realism," in view of its medieval provenance in the explicit recognition of the contrast between *ens reale* and *ens rationis*, I call "hardcore realism," in view of the continuity of the medieval notion of *ens reale* with the ov of which Aristotle spoke as scholasticism's grandfather. Thus, hardcore realism means that there is a dimension to the universe of being which is indifferent to human thought, belief, and desire, such that, for example, if I believe that the soul survives the destruction of my body and I am wrong, when my body goes so does my soul — or, conversely, if I believe that the death of the body is also the end of my mind or soul and I am wrong, when my body disintegrates my soul lives on, and I will have to take stock accordingly. Or, to give a more historical example, from the time of Aristotle to at least that of Copernicus, all the best evidence, arguments, and opinions of every stripe held that the sun moves relative to the earth, while in hardcore reality, all along, supremely indifferent to these stripes of opinion, it was the earth that moved relative to the sun, and the sun also moved, but not relative to the earth.

17. See Poinsot,1632: "First Preamble on Mind-Dependent Being," Article 3, pp. 66/47–72/17.

to the notion of something existing regardless of its status as known. But of course this raises at once the problem: please give an example of something unknown? Well, it is not impossible. What will completely defeat the AIDS virus: that is something just now unknown.[18] The belief that *there* is something, simple or complex, that meets this description is what drives scientific research in the area. When and if it is found the belief will be vindicated, but at that moment this currently "unknown X" will become itself an "object X identified that is not merely an object." So we can say that if something believed in is a part of reality in the hardcore or scholastic sense, then that something has the possibility of passing from unknown to known. At that moment, the only difference will be an extrinsic one: a relation to some knower whereby the thing existing in its own right comes now to exist also as an object, as part of a larger objective world, an item, mayhap, or process therein.

We see thus exactly that and why "such questions as how concepts are related to reality"(Sebeok, 1991b, p. 143) are "ultimately sterile." Reality, for the animal, is simply the objective world, the Umwelt. Later on, but only for the human animal, experience will give rise to the *further* consideration that there seems to be more to objects, a dimension within the objective world, that does not reduce to our experience of objects. How to name this irreducible dimension globally, "generically" — that is to say, without having (or being by any means able!) to specify in detail its specific contents (which is a much more difficult task)? Such is the origin of the *idea* of reality within the human Innenwelt, an idea the medievals termed more expansively "being as first known"(Poinsot, 1633, pp. 24b18-25a31). This idea is a sign, a representation of something other than itself that is largely unknown but determinable and to be determined within experience, something that *distinguishes* the way objects exist in relation to me and my perceptual categories of the "to be sought" (+), "to be avoided" (–), and "the safely ignored" (0) *from* what might be true of those objects apart from such classification ("fere est idem quod cognoscere rem quoad an est" — "it amounts to recognizing that an object does not reduce to my experience of it," as Poinsot says). Thus Sebeok sees the fact that "we can approach the 'real' richness of the universe only by entertaining multiply contending, mutually complementary visions" as but the "quotidian implication" of Niels Bohr's celebrated adage that "physics concerns what we can say about nature"(Pais, 1991, p. 427). I see this quotidian implication as an upshot of the fact that we are inevitably workers in a zoösemiotic context rather than disembodied minds. Bohr (Pais, 1991) is simply wrong, as Sebeok should have been the first to point out, further to conclude *without qualification* that "to think that the task of physics is to find out how nature is" is "wrong." The qualification needed, of course, is the fallibilist one that we can never find out exhaustively "how nature is" — quite another matter than not finding out at all, and Sebeok's real point (Sebeok, 1992a, p. 339) in citing Bohr's statement in the first place.

"Reality" (in the scholastic, or "hardcore," sense), thus, is a representation of objectivity that transcends biological heritage, for it is only indirectly tied to my

---

18. Actually, just now it is not even known that there actually is such a virus to be defeated.

biological type as an organism within a determinate species. Every animal lives in a species-specific objective world, determined from the ground up by its biology. "What is commonly called the 'external world'," in this context, we may be forgiven for considering[19] as "the brain's formal structure (*logos*)" under the stimulus the senses convey from the physical surroundings, as this collusion gives rise to the objective world or Umwelt in which the animal lives and moves and has its being, "models of purlieus frequented by and appropriate to the survival of each organism and its species."[20] This is true also of the human animal, but the uniformity of the objective world is, so to say, partially "ruined" at the level of culture by the incorporation of diverse specifications of "reality" generically considered, as we see in the different customs of marriage, family, and religion (particularly in the matter of which texts — if any, the skeptics will say — have "God himself" or "Allah" or "Jahweh," etc., as their "true author"), not to mention the astronomical controversies which led to the bootless break between medieval coenoscopy and modern ideoscopy. God and the physical environment, thus, are but the polar extremes under the idea of *ens reale* according to which the species-specifically human objective world diversifies itself internally through language exapted in communication.

In other words, the notion of reality is a species-specifically human achievement based on the species-specifically human component of the generically animal modeling system or Innenwelt thanks to which, as Sebeok puts it, we are and cannot but be "workers in a zoösemiotic context;" for we are, even as anthropos, animals from the outset and to the end of our days. We awaken not to a physical environment of pure *ens reale* but to an objective world which, like that of every animal,[21] is a mixture of *ens rationis* and *ens reale* in the presentation and maintenance of objects, the objects we need in order to survive, grow, and flourish. Within these objects what is important is precisely their relation to us, not the "relation to themselves" (itself, note, this "self-relation," an *ens rationis* without which the notion of "hard-core reality" no less than that of "thing" could not arise[22]) which an altruist or, mayhap, a scientifically minded inquirer, might want to pursue.

A Modeling System Biologically Underdetermined

Here Sebeok would typically reveal his modern side by missing a point which his postmodern "better self" was about to make. "It was Niels Bohr who first emphasized the doctrine that scientists have no concern with 'reality'; their job has to do with model building"(Sebeok, 1987a, p. 72). How ironic a point to miss, since the very notion of reality itself results from a modeling within the species-specifically human Umwelt; and further such modeling, both coenoscopic and ideoscopic, aims precisely to clarify the generic intuition in specific ways and circumstances. For, as we have just

---

19. As Sebeok put it, 1992, p. 57.
20. Sebeok ,1995, p. 87 (or 2001, p. 67–68), referencing J. von Uexküll, 1920.
21. See the text of Poinsot 1632, Second Preamble.
22. Guagliardo, 1993; Deely, 1994, Part IV.

seen, "reality," in its philosophical notion as "hardcore" reality, is precisely an achievement, species-specifically human, of modeling, exapted to communicate in the linguistic expression "reality," concern with which — either preclusively (a largely chimerical goal) or as something to be correlatively distinguished within the objective world from the factors of *ens rationis* and identified as such — is equally at the coenoscopic 7th/6th century BC origins of philosophy and the 16th/17th century AD ideoscopic origins of modern science. The concern of scientists, Sebeok aimed at saying, is with the building and testing of models concerned with distinguishing in verifiable ways what in our experience belongs to the order of *ens reale* and what to the order of *ens rationis*, which would not be a problem were it not for the fact that, within human experience, the elements of both orders, however different "in themselves" as connected with subjectivity, are equally objective strands within the semiotic web that we call culture.

So we come by a devious route to what Sebeok called "the ultimate enigma"(Sebeok, 1981a, p. 199): the union of nature and culture within human experience, and the dilemma of disentangling the strands of experience within which this unity is given (first as Umwelt, then, in the wake of the awakening to the idea of "reality," as Lebenswelt, that species-specifically human variant of the generically animal Umwelt insofar as the Umwelt is transformed or modified from within by the components and considerations introduced into the objective horizon of experience by the representative elements formed by language within the Innenwelt and exapted into objective structures of linguistic communication transmogrifying Umwelt to Lebenswelt — all without for a moment suppressing or obviating the "animal roots" of every individual's world as objective, that is to say, experienced and "known" in the context of society as well as individually).

A Glance in the Rear-View Mirror

In my private semiosis, these public considerations carry me back to my youth as a student of philosophy in the school my Dominican professors maintained in River Forest Illinois.[23] My first suspicion of the external world as a quasi-error came not from Sebeok, who gave me the expression, but from Kant, who seemed to me to have imposed on understanding the requirements distinctive rather of sense. I remember visiting the room of one of my professors, Ralph Austin Powell, to inquire whether what Kant had to say of "reason" with it's a-priori forms was not a fundamental confusion of what could only be true of perception insofar as sense is wholly determined biologically by the type of body we happen to have.

"That's a very interesting idea, Brother," Powell replied. "Why don't you write it up as a paper?" The suggestion may have been a device to get me out his room, but in any event it was a good suggestion. It primed me years later to appreciate one of

---

23. The school originally was a Pontifical Faculty, but by my time had also acquired the standard secular accreditation from the North Central agency.

history's greatest ironies in the arena of philosophy. Von Uexküll, by his own attestation, was influenced above all by Kant in arriving at and formulating his *Umweltstheorie*. Kant distinguished only between percepts and concepts, the former arising from sense, the latter from understanding. In fact, this conflation of sensation with perception in the production of representations was a fundamental blunder, for no analysis of knowledge can do without the distinction between sensation prescissively considered as such, wherein mental imagery is superfluously assumed, and perception likewise considered, wherein imagery (or "ideas": *species expressae*, as the Latins generically said[24]), proves essential. But on any such distinction, it becomes quickly apparent that concepts belong to perception before they belong to understanding, and do not belong at all to sensation. Thus the proper question in distinguishing understanding from sense concerns not the difference between sensory "percepts" and rational "concepts," but rather the difference between the concepts proper to understanding and the concepts proper to perception in its difference from sensation.

This last difference is precisely that between concepts of objects classified as to be sought, to be avoided, or to be ignored, and concepts of objects classified as belonging primarily to *ens reale* or *ens rationis*. The relation of an object to itself, which underlies, in the species-specifically human originary grasp of being,[25] the recognition of the difference between objects which are only objects and objects which are also things, is itself already an *ens rationis*, but one consisting in a representation which is not wholly biologically determined, one which therefore belongs to that aspect of the modeling system which Sebeok labels "language" and which is species-specifically human. Communication, but not language, is alone needed for the concepts of perception, and these concepts alone pertain to von Uexküll's "functional cycle," even as the concepts of understanding alone pertain to scientific as distinguished from the ideas of artistic understanding.

In arriving at his concept of the Umwelt, von Uexküll was already employing his animal modeling system in its species-specifically human dimension, which would not have been possible were the Kantian dyadic contrast between "sense" and "reason" the contrast obtaining "in reality." This is why I have said (See Deely, 2001, p. 558n26) that, in a wholly logical world, the study of the purely perceptual intelligence of animals would have been rather the inspiration for the jettisoning of Kantianism in the philosophy of intellectual mind.

Updating the File

The realization that human experience, being animal experience first of all, does not begin simply with *ens reale* but with a world of objects which are normally (at least in historic, if not prehistoric, times) predominantly constituted by *entia rationis* (and

---

24. See the summary in Poinsot 1632: Book II, Question 2, text and notes; and the discussion in the *Four Ages* (Deely, 2001, pp. 345–347).
25. *Ens primum cognitum*, which divides from within, over the course of experience, between *ens reale* and *ens rationis*.

include *entia realia* formally recognized as such only as a virtual dimension and indistinctly as to particulars) is not unprecedented in the history of philosophy.[26] But the full thematization of this realization *is* unprecedented,[27] and may be said to constitute the essence of postmodernity insofar as we are to conceive of it as a distinct philosophical epoch in the wake of the mainstream philosophical development which runs from Descartes in the seventeenth century to Wittgenstein and Husserl in the twentieth. Heidegger pointed to the need for such a thematization under the classical rubric of "being," but he only got as far as the posing of the question[28] to which semiotics begins the answer. Why is it, he asked, in terms with an intersemioticity resonant of von Uexküll, that humans experience beings as present-at-hand, rather than ready-to-hand, which is "closer" to us and indeed the way beings are given proximally and for the most part? The answer lies in the difference, in what is distinctive, of an Umwelt experienced on the basis of an Innenwelt having language as a component in its forming of representations.[29]

The external world is a species specifically human representation. The quasi-error arises from the routine mistaking of objects simply for "things," leading to the confusion of "external reality" (as became the custom within philosophy) with the more fundamental notion of *ens reale*, which is neither identical with "the external world" nor the starting point as such of species-specifically human knowledge, but merely a recognizable dimension experienced within objectivity. The "external world" does not lie beneath or outside of thought and language, as the moderns tended to imagine, but is precisely given, to whatever extent it is given, within objective experience, as semiotics from the first [30] instructed us. Sebeok liked to quote while

---

26. Aquinas, notably (e.g., c.1268/72, *Commentary on Aristotle's Metaphysics*, IV. 6), called it *ens primum cognitum* under which experience leads us to distinguish *reale* from *rationis*: see Deely 2001, pp. 350–57.
27. Guagliardo (1994) is one of the very few who have troubled to unearth some precedents in this move toward thematization.
28. Heidegger, 1927, p. 487. The question was part of the transition to the never completed final sections of this great work.
29. The distinction between the ready-to-hand and the present-at-hand is a distinction that does not arise for any animal except an animal with a modeling system cap-able of representing objects (as such necessarily related to us) according to a being or features not necessarily related to us but obtaining subjectively and/or intersubjectively in the objects themselves (mistakenly or not, according to the particular case) — an animal, in short, capable of *wondering* about things-in-themselves and conducting itself accordingly. Now, since a modeling system so capacities is, according to Sebeok, what is meant by language in the root sense, whereas the exaptation of such a modeling in action gives rise not to language but to linguistic communication, and since 'language' in this derivative sense of linguistic communication is the species-specifically distinctive and dominant modality of communication among humans, we have a difficulty inverse to that of the nonlinguistic animals, although we, unlike they, can overcome the difficulty.

    Our difficulty — the source of the quasi-error of the external world, if I may say so — is that, within an Umwelt, objects *are* reality so far as the organism is concerned. But without language, the animals have no way to go beyond the objective world as such to inquire into the physical environment in its difference from the objective world. Within a Lebenswelt, by contrast, that is to say, within an Umwelt internally transformed by language, the reality so far as the organism is concerned is confused with and mistaken for the world of things. Objects appear not as mixtures of *entia rationis* with *entia realia*, but simply as 'what is', 'real being', 'a world of things'.
30. Using "first" here in the sense of the original treatise which established the unity of signs in the being proper to relation as indifferent to the distinction between *ens reale* and *ens rationis*, namely, the *Tractatus de Signis* of Poinsot, 1632.

constantly reassessing Bohr's asseveration that "We are suspended in language in such a way that we cannot say what is up and what is down"(French & Kennedy, 1985, p.302). In my assessment this is an asseveration whose truth and best interpretation depends on the fact that we are linguistic animals and not just perceptual animals, as I have argued pointedly at some length (Deely, 2002).

As linguistic animals, we can become aware not only of the difference between a thing and an object, between objective world and physical environment as but partially incorporated within objectivity, but we can further become aware of the status of language as a system of signs, and its dependence upon yet other signs in the constitution of objects. It is these objects and their interconnections which go together to form our experience of "reality" (so far like that of any other animal); but within this sphere of objective experience, thanks to language, we can also fashion an *idea* of "reality" establishing an intelligible sense which is not simply given in perception, but is *attained* through sensation *within* perception owing to the difference of sensation from perception.[31] And with that idea thus experimentally grounded, perhaps *only* with that idea, we may say, the human animal begins to awaken to its humanness. Our species is drawn by this aboriginal abduction to set out on the long road of philosophy and science, eventually to come across — quite late in the journey, as it happens — that crossroads having the Way of Signs as one of its forks. At that juncture the human animal realizes that, while every animal and perhaps all nature is *semiosic*, the human animal alone is a *semiotic* animal; and in that moment of realization, which few or none have done more than Sebeok to inaugurate, in philosophy at least, postmodern intellectual culture begins — indeed, takes wing. The quasi-error of the external world need no longer beguile or bemuse us, for its nature and origin have been exposed by the very clearing of the opening to the Way of Signs. We see now to have uncovered a path leading "everywhere in nature, including those domains where humans have never set foot,"[32] but to an understanding of which semiotics gives us the means integrally to aspire. Call it the postmodern interpretive horizon, mayhap, the "coincidence of communication with being"(Petrilli & Ponzio, 2001, p. 54). It is the heart of semiotics, vindicating against modernity the medieval conviction which modern science never wholly abandoned, despite the philosophers: *ens et verum convertuntur*, "communication and being are coextensive." To be for nature is to be intelligible for the animal whose being is to understand.

---

31. Aquinas liked to say that "things are per se sensible but they have to be made intelligible": it is the perpetual task of human understanding in its difference from sensation and perception alike.
32. Emmeche (1994, p. 126); staying silent for the moment on the question over which Sebeok turned conservative, the question of whether semiosis is co-terminus with the emergence of life, or whether there is not indeed a broader origin in which semiosis must be seen as coterminous with the physical universe *tout court*: see Nöth, 2001.

Tom, 9 November 1920–2001 December 21

From the global semiosis which brought about life more than four million years ago to the global semiotics of the twenty-first century, in which a consciousness of that process has begun to be embodied, is the very trajectory that Thomas A. Sebeok himself embodied. He was, if not the first, surely the fullest embodiment of semiotic consciousness so far, since Augustine introduced the thematic possibility of such consciousness with his essay of 397AD. As late as his essay of 1991 (Sebeok, 1991), Tom told us, he was still struggling to make clear "the fact, not then self-evident" — as it had become, for many, by 2001,[33] the last year of Tom's rich and eventful life — "that each and every man, woman, and child superintends over a partially shared pool of signs in which that same monadic being is immersed and must navigate for survival throughout its singular life"(Sebeok, 2001, p. ix).

It is not easy to capture the private side, the Innenwelt, of this most complex of men, who seemed to live in order to build for the Umwelt of public life a new edifice of intellectual culture wherein the human being would finally realize systematically its uniqueness as the only animal able to know that there are signs beyond the making use of and depending upon signs at every level of its life and existence. Organizations and publications sprung up under his hand as tricks from the hand of a magician. The more than twenty years of annual *Proceedings* volumes of the Semiotic Society of America, in particular, provide an intellectual record of a community of inquirers drawn together by Sebeok's genius in an organizational meeting of 1975,[34] which resulted in the incorporation under its Constitution for the first formal Annual Meeting in Atlanta the following year[35] of that organization.

Between 1976 and the 2000 twenty-fifth Annual Meeting under the theme of "Sebeok's Century,"[36] the number of semiotic meetings quite apart from American semiotics organized by or around Sebeok would not be easy to enumerate. When, in Imatra, Finland, in the summer of 2000, Sebeok announced to his incredulous audience at Tarasti's annual International Summer Institute that this would be his last

---

33. The medieval Latins commonly distinguished two kinds of propositions under the heading of "self-evident" (*per se nota or selbstverständlichkeit*), namely, those self-evident to anyone understanding the immediate sense of the terms themselves from which the proposition is formed (*propositiones per se nota quoad omnes*), and those self-evident only "to the wise," i.e., to those who understand not merely the terms as such but the further implications that follow from their arrangement in this particular proposition, who have achieved a grasp of the larger context of intelligibility within which the proposition in question is able to maintain its sense (*propositions per se not quoad sapientes*). Tom is saying, by way of Introduction to the final book completed within his lifetime, that the proposition that human experience throughout is an irreducible, labile interweave of sign-relations both mind-dependent and mind-independent, is a proposition that has become self-evident within semiotics by the time we have entered the twenty-first century, a *propositio per se nota quoad sapientes*, something self-evident to semioticians insofar as they have come to understood that the being proper to signs consists in triadic relations indifferently real and unreal according to circumstance.
34. The first North American semiotics colloquium held in Tampa, July 28–30, 1975, at the University of South Florida, and memorialized in Sebeok, Ed., 1977.
35. Of this inaugural meeting, an informal partial proceedings privately edited and published by Charls Pearson survives.
36. See the Editor's Preface of that title in the *Semiotics 2000* Proceedings volume.

visit to Europe, perhaps he, who told Susan Petrilli by telephone in his final weeks that "It is very boring to die," had some secret premonition of the end no one else could think of as near at hand. Yet only about eighteen months remained.

Caption 1: View from beside the Podium of Sebeok,s Last Imatra Talk; audience includes Ttraian Stanciulescu, Christina Ljundberg, Sören Brier, Solomon Marcus, Marcel Danesi, Eila Tarasti, Dines Johansen, John Deely, Eero Tarasti, and Vilmos Voigt.

I think back over the years, over many occasions, occasions that call for volumes to be written, many volumes, and from many points of view. But I will choose to narrate only from one such occasion a little story to close this melancholy. It was the opening day of the 1990 Hungaro-Austrian conference convened to mark the occasion of Tom's seventieth birthday (See Bernard et al., 1990). The day was to close with a wine and cheese gathering in the early evening, after the last paper, and for the occasion the participants were quietly invited each to prepare a toast in Tom's honor, and at the reception we planned to surprise him with a round of prepared toasts.

Unfortunately for that day, the last scheduled speaker was not able to attend the conference, but had sent a student to read her paper, in French, with explicit instructions to "leave nothing out." Well, the paper was far too long for the time frame, but the student felt bound by her instructions, and the chair of the session was perhaps too intimidated by the status of the author in Tom's circle to intervene. It is likely that this chair was unfamiliar with Tom's own example in such matters, set at the Tampa meeting mentioned above, when he as session chair simply stepped in between the offending speaker and the microphone to introduce the next speaker when the offending speaker's allotted time expired — a story I reserve for another occasion.[37]

---

37. I wrote the incident up in Bari, Italy, February 19, 2002, under the title, "Tom Sebeok, the Man Who Loved Time," for the disposition of Jean Umiker-Sebeok.

Caption 2: Tom Sebeok with his daughters Jessica and Erica in Budapest on the opening day of the Hungaro-Austrian conference celebrating his approaching 70th birthday.

The poor student droned on and on. By the time the session closed, the time allotted for the wine and cheese reception had so truncated that the round of toasts was quietly canceled by the organizers.

I have no idea how many of the participants had prepared toasts, nor whether Tom ever knew of the planned round. My own toast, undelivered at the reception, we later used to dedicate the Budapest-Vienna volume in which the conference eventuated (Bernard et al., 1990, p. iv) but, changing present tenses as appropriate, I would like to repeat that toast here to close the present essay around Tom's work in *memoriam*:

> *Dr. Sebeok was a man of extraordinary talents,*
>   *we all know. That in itself is*
>   *not extraordinary.*
> *What was extraordinary was what was beneath the talents,*
>   *namely, the way they were orchestrated.*
> *And how was that?*
> *Dr. Sebeok, somehow,*
>   *so directed the play of his own talents that*
>   *the talents of all who associated with him*
>   *were also brought into play.*
> *He managed in this way*
>   *to bring a thousand and more than a thousand individuals*
>   *who would otherwise have never known one another*
>   *into a kind of intellectual symphony or orchestra*
>   *whose works collectively express*
>   *— through his direction —*

*most of what is best
in that movement we call today "semiotics"
We celebrate in this volume
what was extraordinary in Dr. Sebeok,
what the ancient Greeks and medieval Latins would call
his ψυχη, anima, or "soul."
May it live forever!*

It will in our hearts, and in the life of human culture. "An academic," Tom averred on the occasion of his transition to Emeritus status in 1991,[38] "is the sign's way of spawning further, more developed academics." To accomplish this, he went on to say, "there are two fundamental strategies." First, one must publish and teach "as much as possible;" second, "equally important," one must do one's best "to facilitate the success of one's colleagues in these respects." Successful execution of these two fundamental strategies, he averred, "are the only things I have ever wanted to do in my academic life." He did both splendidly. Not even death cancels the achievement.

REFERENCES

AQUINAS, Thomas (1224/5–1274).
   i.1252–1273. *S. Thomae Aquinatis Opera Omnia ut sunt in indice thomistico,* ed. Roberto Busa (Stuttgart-Bad Cannstatt: Frommann-Holzboog, 1980), in septem voluminas.
   c.1268/72. *In duodecim libros metaphysicorum Aristotelis expositio,* in Busa ed.vol. 4, 390–507.
ASHLEY, Benedict.
   1985. *Theologies of the Body: Humanist and Christian* (Braintree, MA: The Pope JohnXXIII Research Center, 1985.
AUGUSTINE of Hippo (354–430AD).
   i.397–426AD. *De doctrina christiana libri quattuor* ("Four Books On Christian Doctrine"), in Tomus Tertius Pars Prior, pp. 13–151, of *Sancti Aurelii Augustini Hipponensis Episcopi Opera Omnia,* opera et studio Monachorum Ordinis Sancti Benedicti e congregatione S. Mauri (Ed. Parisina altera, emen-data et aucta: Gaume Fratres, 1836; copy at the Xochimilco Dominican priory in Mexico City); also in *Patrologiae Cursus Completus,* ed. J. P. Migne, *Series Latina* (PL), Volume 34, cols. 15-122.
BERNARD, Jeff, John DEELY, Vilmos VOIGT, and Gloria WITHAL, Editors.
   1990. *"Symbolicity"*. Papers from the International Semioticians' Conference in Honor of Thomas A. Sebeok's 70th Birthday Budapest-Wien, September 30–4 October 1990 (= Sources in Semiotics, XI, bound together with *Semiotics 1990;* Lanham, MD: University Press of America, 1993).
DEELY, John.
   1980. "The Nonverbal Inlay in Linguistic Communication", in *The Signifying Animal,* ed. Irmengard Rauch and Gerald F. Carr (Bloomington, IN: Indiana University Press), pp. 201–217.
   1984. "Semiotic as Framework and Direction", presented in the "Semiotic: Field or Discipline?" State-of-the-Art Conference organized by Michael Herzfeld at Indiana University (Bloomington) 8–10 October; published in Deely, Williams, and Kruse 1986: 264–271.
   1985. "Editorial AfterWord" and critical apparatus to *Tractatus de Signis: The Semiotic of John Poinsot* (Berkeley: University of California Press), 391–514; electronic version hypertext-linked (Charlottesville, VA: Intelex Corp.; see entry under Poinsot 1632a below).
   1986c. "A Context for Narrative Universals. Semiology as a *Pars Semiotica*", *The American Journal of Semiotics* 4.3-4, 53-68.
   1993a. "How Does Semiosis Effect Renvoi?", the Thomas A. Sebeok Fellowship Inaugural Lecture delivered at the 18th Annual Meeting of the Semiotic Society of America, October 22, 1993, St. Louis, MO; published in *The American Journal of Semiotics* 11.1/2 (1994), 11–61; text available also as Ch. 8 of Deely 1994a: 201–244.
   1994. *The Human Use of Signs; or Elements of Anthroposemiosis* (Lanham, MD: Rowman & Littlefield).

---

38. Sebeok 1992b: his "retirement" speech, "Into the Rose Garden," delivered on March 22, 1991, but published only the following year.

2001. *Four Ages of Understanding. The first postmodern history of philosophy from ancient times to the turn of the 21st century* (Toronto, Canada: University of Toronto Press).
2001e. "A Sign is *What*?", *Sign Systems Studies* 29.2
2001f. "Umwelt", *Semiotica* 134–1/4, 125–135.
2002. *What Distinguishes Human Understanding?* (South Bend, IN: St. Augustine's Press).
2003. *The Impact on Philosophy: Semiotics and the Error of the External World, with a Dialogue between a Semiotist and a Realist* (South Bend, IN: St. Augustine's Press).

DEELY, John N., Brooke WILLIAMS, and Felicia E. KRUSE, editors.
1986. *Frontiers in Semiotics* (Bloomington: Indiana University Press). Preface titled "Pars Pro Toto", pp. viii–xvii; "Description of Contributions", pp. xviii–xxii.

EMMECHE, Claus.
1994. *The Garden in the Machine* (Princeton, NJ: Princeton University Press).

FRENCH, Anthony Philip, and KENNEDY, P. J., Editors.
1985. *Niels Bohr, A centenary volume* (Cambridge, MA: Harvard University Press).

GUAGLIARDO, Vincent (1944–1995).
1993. "Being and Anthroposemiotics", in *Semiotics 1993*, ed. Robert Corring-ton and John Deely (Lanham, MD: University Press of America, 1994), 50–56.
1994. "Being-as-First-Known in Poinsot: A-Priori or Aporia?", *American Catholic Philosophical Quarterly* 68.3 "Special Issue on John Poinsot" (Summer), pp. 363–393.

HEIDEGGER, Martin (1889–1976).
1927. *Sein und Zeit, originally published in the Jahrbuch für Phänomenologie und phänomenologische Forschung*, ed. E. Husserl. Page references in the present work are to the 10th edition (Tübingen: Niemeyer, 1963).

JACOB, François.
1982. *The Possible and the Actual* (Seattle, WA: University of Washington Press).

JAKOBSON, Roman (1896–1982).
1974. "Coup d'oeil sur le devéloppement de la sémiotique", in *Panorama sémiotique/A Semiotic Landscape,* Proceedings of the First Congress of the International Association for Semiotic Studies, Milan, June 1974, ed. Seymour Chatman, Umberto Eco, and Jean-Marie Klinkenberg (The Hague: Mouton, 1979), 3-18. Also published separately under the same title by the Research Center for Language and Semiotic Studies as a small monograph (= Studies in Semiotics 3; Bloomington: Indiana University Publications, 1975); and in an English trans. by Patricia Baudoin titled "A Glance at the Development of Semiotics", in *The Framework of Language* (Ann Arbor, MI: Michigan Studies in the Humanities, Horace R. Rackham School of Graduate Studies, 1980), 1–30.

KULL, Kalevi, Guest-Editor.
2001. *Jakob von Uexküll: A Paradigm for Biology and Semiotics*, a Special Issue of *Semiotica* 134–1/4.

LANGER, Susanne K. (1895–1985).
1971. "The Great Shift: Instinct to Intuition", in J. F. Eisenberg and W. S. Dillon, Eds., *Man and Beast: Comparative Social Behavior* (Smithsonian Annual III; Washington, DC: Smithsonian Institution Press), 313–332.

LYOTARD, Jean-François.
1984. *The Postmodern Condition: A Report on Knowledge* (Minneapolis, MN: University of Minnesota Press), trans. Geoff Bennington and Brian Massumi of La Condition postmoderne: rapport sur le savoir (Paris: Editions Minuit, 1971).

NÖTH, Winfried, Organizer.
2001. German-Italian Colloquium "The Semiotic Threshold from Nature to Culture", Kassell, Germany, 16–17 February at the Center for Cultural Studies, University of Kassel; papers published together with the Imatra 2000 Ecosemiotics colloquium in *The Semiotics of Nature*, a Special Issue of *Sign System Studies* 29.1, ed. Kalevi Kull and Winfried Nöth. (This journal, founded by Juri Lotman in 1967, is the oldest contemporary journal of semiotics, and, interestingly, appeared in its first three issues under the original version of Locke's coinage, Σημιωτικη rather than Σημειωτικη: see the extended discussion of the matter in the Four Ages Chapter 14.)

PAIS, Abraham.
1991. *Niels Bohr's Times, in Physics, Philosophy, and Polity* (Oxford: Clarendon Press).

PEIRCE, Charles Sanders (1838–1914).
Note. The designation CP abbreviates *The Collected Papers of Charles Sanders Peirce,* Vols. I–VI ed. Charles Hartshorne and Paul Weiss (Cambridge, MA: Harvard University Press, 1931–1935), Vols. VII–VIII ed. Arthur W. Burks (same publisher, 1958); all eight vols. in electronic form ed. John Deely (Charlottesville, VA: Intelex Corporation, 1994). Dating within the CP (which covers the period in Peirce's life i.1866–1913) is based principally on the Burks Bibliography at the end of CP 8. The abbreviation followed by volume and paragraph numbers with a period between follows the standard CP reference form.
1905. "What Pragmatism Is", *The Monist* 15 (April), 161–181; reprinted in CP 5.411–437, with 5.414–5.435 being editorially headed "Pragmaticism" in CP.

PETRILLI, Susan, and PONZIO, Augusto.
- 2001. *Thomas Sebeok and the Signs of Life* (USA: Totem Books).

POINSOT, John.
- 1632. *Tractatus de Signis*, subtitled *The Semiotic of John Poinsot*, extracted from the *Artis Logicae Prima et Secunda Pars* of 1631–1632 using the text of the emended second impression (1932) of the 1930 Reiser edition (Turin: Marietti), arranged in bilingual format by John Deely in consultation with Ralph A. Powell (First Edition; Berkeley: University of California Press, 1985). This work, the first systematic treatise on the foundations of semiotic, is also available as a text database, stand-alone on floppy disk or combined with an Aquinas database, as an Intelex Electronic Edition (Charlottsville, VA: Intelex Corp., 1992).
- 1633. *Naturalis Philosophiae Prima Pars* (Madrid, Spain). In Reiser vol. II: 1–529.

RUSSELL, Bertrand (1872–1970).
- 1959. *My Philosophical Development* (New York: Simon and Schuster).

SEBEOK, Thomas A.
- 1976. *Contributions to the Doctrine of Signs* (=Sources in Semiotics IV; Lanham, MD: University Press of America, 1985 reprint with a new Preface by Brooke Williams of the original book as published by Indiana University, Bloomington, and The Peter De Ridder Press, Lisse).
- 1977a. "Ecumenicalism in Semiotics", in *A Perfusion of Signs*, ed. Sebeok (Bloomington, IN: Indiana University Press, 1986), 180–206.
- 1981a. *The Play of Musement* (Bloomington, IN: Indiana University Press).
- 1984. "Vital Signs", Presidential Address delivered October 12 to the ninth Annual Meeting of the Semiotic Society of America, Bloomington, Indiana, October 11–14; subsequently printed in *The American Journal of Semiotics* 3.3, 1–27, and reprinted in Sebeok 1986: 59–79.
- 1986. *I Think I Am A Verb. More Contributions to the Doctrine of Signs* (New York: Plenum Press).
- 1986a. "Toward a Natural History of Language", in *The World & I* (October 1986), 462–469, as reprinted in Sebeok 1991a: 68–82, to which reprint pagination references in this essay are keyed.
- 1987. "Language: How Primary a Modeling System?", in *Semiotics 1987*, ed. John Deely (Lanham, MD: University Press of America, 1988), 15–27.
- 1987a. "Toward a Natural History of Language", *Semiotica* 65, 343–358 (expanded from a review article in *The World & I* of October 1986, pp. 462–469), as reprinted in Sebeok 1991a: 68–82.
- 1991. *Semiotics in the United States* (Bloomington, IN: Indiana University Press).
- 1991a. *A Sign is Just a Sign* (Bloomington, IN: Indiana University Press).
- 1991b. "Indexicality", in Sebeok 1991a: 128–143.
- 1992. "Galen in Medical Semiotics", as finally published in Sebeok 2001: 44–58 (see the bibliographical note on p. 44 bottom).
- 1992a. "'Tell Me, Where Is Fancy Bred?': The Biosemiotic Self", in *The Semiotic Web* 1991, ed. T. A. Sebeok and J. Umiker-Sebeok (Berlin: Mouton de Gruyter), 333–343 (reprinted also in Sebeok 2001: 120–127).
- 1992b. "Into the Rose-Garden", *Ural-Altaische Jahrbücher* 64, 1–12.
- 1995. "Semiotics as Bridge between Humanities and Sciences", in the volume of this same title ed. Paul Perron, Leonard G. Sbrocchi, Paul Colilli, and Marcel Danesi (Ottawa, Canada: Legas, 2000), 76–100. Reprinted under the title "Signs, Bridges, Origins", in Sebeok 2001: 59–73, with the addition of a brief concluding eight-line "Erato's Coda".
- 2000. "Some Reflections on Vico in Semiotics", in *Functional Approaches to Language, Culture and Cognition*, ed. D. G. Lockwood, P. H. Fries, and J. E. Copeland (Amsterdam: John Benjamins), 555–568; as reprinted in Sebeok 2001: 135–144, to which reprint page references in this essay are keyed.
- 2001. *Global Semiotics* (Bloomington, IN: Indiana University Press).
- 2001a. "Biosemiotics: Its Roots, Proliferation, and Prospects", in Kull Guest-Ed. 2001: 61–78.

SEBEOK, Thomas A., Editor.
- 1977. *A Perfusion of Signs. Transactions of the First North American Semiotics Colloquium* (Bloomington, IN: Indiana University Press, 1986), 180–206.

SIMPKINS, Scott, and John DEELY, Editors.
- 2000. *Semiotics 2000: "Sebeok's Century"* (25th Annual Proceedings Volume of the Semiotic Society of America; Ottawa, Canada: Legas Publishing, 2001).

SOKAL, Alan, and Jean BRICMONT.
- 1998. *Fashionable Nonsense. Postmodern intellectuals abuse of science* (New York: Picador USA).

von UEXKÜLL, Jakob.
- 1928. *Theoretische Biologie* (Berlin; 2nd ed.; reprinted Frankfurt a. M.: Suhrkamp 1970).

WOJTYLA, Karol Jósef.
- 1998, September 14. *Fides et Ratio*, encyclical letter on the relationship between faith and reason (Rome, Italy: Vatican City).

Thomas A. Sebeok and biology:
Building biosemiotics

Kalevi Kull[1]

Abstract: The paper attempts to review the impact of Thomas A. Sebeok (1920–2001) on biosemiotics, or semiotic biology, including both his work as a theoretician in the field and his activity in organising, publishing, and communicating. The major points of his work in the field of biosemiotics concern the establishing of zoosemiotics, interpretation and development of Jakob v. Uexküll's and Heini Hediger's ideas, typological and comparative study of semiotic phenomena in living organisms, evolution of semiosis, the coincidence of semiosphere and biosphere, research on the history of biosemiotics.

Keywords: semiotic biology, zoosemiotics, endosemiotics, biosemiotic paradigm, semiosphere, biocommunication, theoretical biology

> "Culture," so-called, is implanted in nature; the
> environment, or Umwelt, is a model generated
> by the organism. Semiosis links them.
> T. A. Sebeok (2001c, p. vii)

When an organic body is dead, it does not carry images any more. This is a general feature that distinguishes complex forms of life from non-life. The images of the organism and of its images, however, can be carried then by other, living bodies. The images are singular categories, which means that they are individual in principle. The identity of organic images cannot be of mathematical type, because it is based on the recognition of similar forms and not on the sameness. The organic identity is, therefore, again categorical, i.e., singular.

Thus, in order to understand the nature of images, we need to know what life is, we need biology — a biology that can deal with phenomena of representation, recognition, categorisation, communication, and meaning. This is a special kind of biology, richer than the one built according to the rules of the methodology of natural science. A powerful contribution to such extended general biology has been made by Thomas A. Sebeok. The following words can be found in Winfried Nöth's *Handbuch der Semiotik*:

> Als Pionier der Semiotik des 20. Jahrhunderts verdient Thomas Albert Sebeok (geb. 1920[2]) besondere Erwähnung. ... Sebeok hat sich durch umfangreiche editorische Tätigkeiten um die internationale Verbreitung der Semiotik Verdienste erworben. ... In seinen eigenen Arbeiten zur

---

1. Department of Semiotics, University of Tartu, Tiigi St. 78, 50410 Tartu, Estonia; kalevi@zbi.ee
2. Born on November 9, 1920, in Budapest, Hungary.

> Semiotik … plädiert Sebeok für die Erweiterung der Semiotik und die Überbrückung der Grenzen zwischen den Geistes- und Naturwissenschaften im Rahmen der Semiotik. Die Entstehung und Entwicklung der Zoosemiotik, der Biosemiotik und der Evolutionären Semiotik als neue Teilgebiete der Semiotik in Erweiterung der Anthroposemiotik sind wesentlich mit dem Namen T. A. Sebeoks verbunden. (Nöth, 2000, pp. 42–43)

On December 21, 2001, T. A. Sebeok died in Bloomington (Indiana, USA), the city where he lived and worked most of his life. As a great designer of semiotics, his importance is far more fundamental than can be described here, or in any single article. A good minimal account of him has been collected in an obituary by J. Bernard (2002), in addition to other recent obituaries (Hoffmeyer, 2002; Kull et al., 2002; Petrilli, 2002; etc.), numerous writings from Sebeok's lifetime (Baer, 1987; Deely, 1995a, 1998; Danesi 2000, 2001; Nuessel, 2000; Petrilli & Ponzio, 2001; Ponzio & Petrilli, 2002; etc.), and large collective *festschrifts* (Bouissac et al., 1986; Bernard et al., 1993; Tasca, 1995; Tarasti, 2000). Almost all of these, at least to some extent, mention Sebeok's work in relation to biology. As E. Baer (1987, p. 182) has said, "the point of departure for Sebeok's doctrine of signs is found in biology." However, until now there does not exist, according to my knowledge, any writing that would try to review his biological work. Let me put the latter as the aim of the current writing. And since, for Sebeok, the scholarly research was always intertwined by developing the web between scholars, this aspect will also be reflected here.

Thus, on one hand, this paper unintentionally belongs to a series of studies that we have planned together with Tom Sebeok, about the classical figures whose work has been important for the formation and development of biosemiotics, or semiotic biology.[3] On the other hand, I want to stress here that Sebeok's work described below belongs to true biology, it is about the foundations of biology, which is more than an application of a semiotic approach in certain aspects of biology or an analysis of biological aspects of semiotics. This is an extension of biology beyond the natural science, beyond a subjectless biology. Actually, an evident step that had to be taken anyway, in order to understand life and not just to describe it.

Synopsis

The work and impact of Thomas A. Sebeok on the development of biosemiotics require a special volume, because studying his works will be a necessary part of education for everybody who wants to inquire into the semiotic basis of life science. Here, I will list very briefly a few points of his foundational work in this field.

---

3. This series already includes publications on semiotics classics in their relationship to biology, exempli gratia, on Ch. S. Peirce (Santaella, 1999), Ch. Morris (Petrilli, 1999b), R. Jakobson (Shintani, 1999), J. Lotman (Kull, 1999b), V. Welby (Petrilli, 1999a), as well as on biologists and others who have made a remarkable impact for biosemiotics, as on J. v. Uexküll (Kull, 2001), G. Prodi (Cimatti, 2000), H. Hediger (Turovski, 2000; Sebeok, 2001b; see a review of the latter in Carmeli, 2002), F. S. Rothschild (Kull, 1999c), G. Bateson (Brauckmann, 2000), G. E. Hutchinson (Anderson, 2000).

Much of Sebeok's effort has been concentrated on one central question: "whether a truly comparative science of signs is possible" (Sebeok, 1972, p. 1). In the context of semiotic biology, the following points in Sebeok's work should be emphasised:

(a) Establishing zoosemiotics. Sebeok is the author of the term 'zoosemiotics' (from 1963), and he has published widely on the problems of animal communication. This includes the compiling of zoosemiotic bibliography (Sebeok, 1969), numerous papers and books in the field (Sebeok, 1963, 1969, 1972, 1990), and the editing of large volumes of collective works on zoosemiotics (Sebeok, 1968; Sebeok & Ramsay, 1969; Sebeok & Umiker-Sebeok, 1980; Sebeok & Rosenthal, 1981).
(b) Analysing the basic sign types in their applicability and use by non-human organisms (e.g., Sebeok, 1977, 1991).
(c) Introducing the endosemiotic sphere — signs in the body — as different from zoosemiotics (Sebeok, 1976).
(d) Analysing the concept of biosemiotic self (Sebeok, 1992).
(e) Discussing Lotman's typology of sign systems, and arguing for the existence of primary modelling systems as those of the pre-linguistic or non-verbal ones; then, the linguistic modelling systems will be the secondary ones (Sebeok, 1994, 1996b).
(f) Discussing on Lotman's concept of semiosphere, and arguing for the inclusion of non-human sign systems into it (Sebeok, 2000); i.e., broadening the scope of semiotics to include the biosphere (Sebeok, 2002).
(g) Introducing the methods of semiotic analysis for biosemiotic systems (Sebeok & Danesi, 2000).
(h) Organising, supporting, and editing many collective works on biosemiotics (e.g., Sebeok & Umiker-Sebeok, 1992).
(i) Working on the history of biosemiotics. This includes the writings about Jakob von Uexküll (1864–1944) (Sebeok 1977, 1998), Heini Hediger (1908–1992) (Sebeok, 2001b), and framing the history of biosemiotics in general (Sebeok, 1996a, 1999a, 2001a).

Below, these points will be described in few more details.

Zoosemiotics

Sebeok started his scientific work as a Finno-Ugric linguist, coming from Hungary. Among his major teachers were Charles Morris in Chicago and Roman Jakobson at Princeton.[4] Trying to trace the signs of his movement towards biology, one can mark his early interest in general and interdisciplinary problems. For instance, his paper together with Giuliano Bonfante (Bonfante & Sebeok, 1944) argued for the applicability of a (originally) biological 'age and area' hypothesis (or Willis' law, according to English plant geographer John Cristopher Willis who has described this

---

4. Both are also mentioned by Sebeok (2001c, p. 3) as directing the attention of semiotics towards biology.

rule in his book of 1922) in linguistics — of course, with interesting exceptions. After 1954, Sebeok also wrote on psycholinguistics, where some of his zoological interests are seemingly rooted,[5] but the first appearance of directly zoological topic dates only to 1962 (Sebeok 1962).[6] Since then, animal communication has become a frequent topic of his publications (Sebeok, 1963, 1965a, 1968). Most of Sebeok's publications in the field from this first decade have been included in his book *Perspectives in Zoosemiotics* (Sebeok 1972).

At first, his interest turned to the study of codes in animal communication (Sebeok 1962, 1965c). According to his definition, "by code is meant everything that the source and the receiver know *a priori* about the message" (Sebeok, 1972, p. 9). Therein, one of the questions he paid attention to was the relationship between analog and digital coding.[7] Sebeok developed "the hypothesis that whereas subhuman species communicate by signs that appear to be most often coded analogically, in speech ... some information is coded [analogically] and other information is coded digitally" (Sebeok, 1972, p. 10). Sebeok's interest clearly reflects the general influence that the developing fields of cybernetics and information theory had on linguistics of that period. These, altogether, led to interdisciplinary communication studies in animals and men.[8]

About the same time when zoology started to be his field, he also enters the field of semiotics.[9] The remarkable fact that these turns were closely related for Sebeok clearly helps in understanding his thinking.

Quite soon after that, he started to use the term 'zoosemiotics' (Sebeok, 1965b). Most probably, this term was first coined by him (in Sebeok, 1963, p. 465).[10] He started to pay attention to the relationship between ethology and semiotics. He tried to review the field of animal communication research, compiling a bibliography of the field and publishing it in several versions (Sebeok, 1969, pp. 210-231, 1972, pp. 134-161). He could indeed collect an amazingly rich library on animal communication studies. Sebeok provides many examples of sign use in animals, and classifies them on the basis of sign types. He tends to claim that the decisive role in animal behaviour belongs to indexical signs: "The survival of all species, and of each individual member of every species, depends on the correct decipherment of indexical signs ceaselessly barraging their Umwelt" (Sebeok, 1997b, p. 282).

---

5. About this first period of research, see his own description in Sebeok, 1986a, pp. ix-xi, 65; 1995. Cf. Baer, 1987, p. 181.
6. There exists a comprehensive bibliography of Sebeok's writings of 1942–1995, published by John Deely (1995b).
7. It is interesting to mention in this respect that one of the first works of Danish biosemioticians Jesper Hoffmeyer and Claus Emmeche (1991) was devoted to the same problem.
8. An expression by Hans Kalmus (1906–1989) may illustrate this: "Nevertheless no organism, solitary or social, is conceivable, which has not grown up under the control of a well-integrated communication system, the element of which are the genes" (Kalmus 1950: 22; see also Kalmus 1962).
9. "By 1962, I had edged my way into animal communication studies. Two years after that, I first whiffled through what Gavin Ewart evocatively called 'the tulgey wood of semiotics'" (Sebeok, 1986a, p. ix).
10. A detailed story can be found in the chapter "The word 'zoosemiotics'" in Sebeok, 1972, pp. 178–181.

Then, he enters into a discussion on the existence of language in animals, denying it on the basis of an analysis of the example of Wilhelm von Osten's trained horse Kluge Hans, which was studied already by Oskar Pfungst (Sebeok, 1980; Umiker-Sebeok & Sebeok, 1980; Sebeok & Rosenthal, 1981). The period coincided with an intensified work on teaching language to human apes, and Sebeok began to be strongly critical towards these approaches which were blind to the categorical difference between language and animal communication.[11]

Sebeok's position in using the term 'language' was very clear: "Expressions such as 'language of the bees', even when used with the authority of a Nobel Laureate, Karl von Frisch, are metaphors;" "picturesque combinations of the word 'language' with the generic word 'animal' ... ape or dolphin, or a category of domestic pets (cat, dog), or in phrases like 'the language of flowers', are unscientific nonsense, examples of *petitio principii*" (Sebeok, 1996b, pp. 105–106). Another statement defines the difference:

> All the animals paleontologists classify generically as *Homo*, and only such, embody, in addition to a primary modelling system ..., a secondary modelling system, equivalent to a natural language. The difference amounts to this: while the Umwelten of other animals model solely a (for each) 'existent world', man can, by means of the secondary system, also model a potentially limitless variety of 'possible worlds' (containing sentences with alethic, deontic, or epistemic modalities). (Sebeok, 1996b, p. 106)

Despite the great influence Sebeok's works have had on the study of semiotics of animal communication (and on linguistics and biology; see, e.g., Smith, 1974; Ruse, 1998), the responses he personally received from the specialists in the field were not always satisfactory to him. Those who worked in ethology (mostly within the neo-Darwinian paradigm), did not see the zoosemiotic approach as operational enough. And those who studied the linguistic behaviour of apes thought that Sebeok's critique had not been entirely to the point. This has probably been an additional reason for his search for more fundamental principles of biosemiotics.

Biosemiotics

The step Sebeok was able to make from zoosemiotics to biosemiotics has quite evidently been a result of, on one hand, reading the classical works of Jakob von Uexküll at the end of 1970s, and on the other hand his conversations with Thure von Uexküll and Giorgio Prodi. He has himself described the details of these meetings on several occasions (e.g., Sebeok, 1998). This turn also had a Russian dimension, via a book by Stepanov (1971) which he came across probably soon after its publication and which opens with a chapter titled 'Biosemiotics'. However, Sebeok himself hesitated to use this term for a long time. For instance, the collective paper that

---

11. In Sebeok (1986a, pp. 189-213), one can find the reprintings of his reviews on the works of Rumbaughs, Premacks, and others who attempted to teach human language to apes. These discussions are reminded until today (e.g., O'Connor, 2002).

appeared in *Semiotica* in 1984 (Anderson et al., 1984)[12] and has formulated a direct research program for semiotic biology, still avoided this term, as well as his dictionary of 1986 (Sebeok, 1986b).[13]

In a way, the turn toward biosemiotics has probably something to do with changes in general semiotics. This becomes clear when semiotics of the 1960s and 70s is compared to semiotics in the 1990s. For instance, if in the first period Roman Jakobson's influence was considerable, then in the second period an emphasis on the theoretical concepts of Charles Peirce became a dominating one. This also means a change in the central concepts, from message, sender, and receiver, to sign (or text), semiosis, and interpretant.

Since 1977, Sebeok became interested in the concept of "the semiotic self" (Sebeok, 1986a, p. xi, 1992, p. 335). This includes a problem of "how are self-images established, maintained, and transmuted into performances" (Sebeok, 1992, p. 334). He pointed out that "bodily sensations and the like, most saliently among them those connected with illness, are not amenable to verbal expression because they lack external referents" (Sebeok, 1992, p. 336). He proposed "to discriminate between two apprehensions of the self, (a) the immunologic or biochemical self, with, however, semiotic overtones, and (b) the semiotic or social self, with, however, biological anchoring," thus showing that "the self is a joint product of both natural and cultural processes" (Sebeok, 1986a, p. xi).

The problem of semiotic self is inherently related to the notion of endosemiosis — a field introduced very much due to Sebeok (in Sebeok 1976, this concept has been proposed; see also Sebeok, 2001c, p. 20).

There has been a well-known debate about the concepts of primary and secondary modelling systems (see, e.g., Sebeok & Danesi, 2000). According to the initial formulation by Lotman, language is the primary modelling system, whereas culture comprises the secondary one. Later, Sebeok argued that there exists the zoosemiotic system which has to be called the primary one, leaving the secondary status to language, and the tertiary one to culture (e.g., Sebeok, 1994). Sebeok's view has been supported by many later authors (cf. Moriarty, 1994).

T. A. Sebeok, who has argued for introducing semiotics into all areas of biology, has found it reasonable to specify the terms in corresponding ways. All main types of living creatures serve as an object for semiotic analysis:

> According to one standard scheme for the broad classification of organisms, five superkingdoms are now distinguished: protists; bacteria; plants; animals; and fungi. In each group, distinct but intertwined modes of semiosis have evolved. (Sebeok,1997a, p. 440)

---

12. The writing of this *manifesto* has been proposed by Sebeok. The drafts were written by M. Anderson and circulated for comments and additions among other authors. About more details on the formation of this paper see Sebeok, 1986a, pp. 17–18.
13. I also remember how curiously Sebeok questioned me about the term 'biosemiotics' when I freely used it during my talk at the Glottertal meeting, 1992. As we learned much later, the term had been used already in 1962 by F. S. Rothschild (Kull, 1999c).

Indeed, as the first major distinction is into kingdoms, and biology is using corresponding divisions in scientific inquiry as bacteriology, protistology, botany, mycology, and zoology, one can, correspondingly, apply biosemiotic divisions for each kingdom — e.g., *bacteriosemiotics, phytosemiotics, mycosemiotics, zoosemiotics,* etc. Such a terminology would emphasize that there exist two principal ways in studying the organisms — (a) on the basis of a methodology of natural science, and (b) on the basis of an extended — semiotic — methodology, which is the methodology of the sciences of meaning (*Bedeutungswissenschaften*).

Sebeok, whose particular emphasis is on the plant/animal/fungus trichotomy, does not take these categories as levels, but more as the complementary ones:

> These three categories, distinguished by taxonomers according to the nutritional patterns of each class, that is, three different ways in which information (negentropy) is maintained by extracting order out of their environment, are complementary. (Sebeok, 1997a, p. 441)

He also notes, but does not explore, "the remarkable parallelism between this systematists' P-A-F [plant-animal-fungus] model and the classic semioticians' O-S-I [object-sign-interpretant] model" (Sebeok, 1997a, p. 441). This is because "on this macroscopic scale animals can be catalogued as intermediate transforming agents between two polar opposite lifeforms: the composers, or organisms that 'build up', and the decomposers, or organisms that 'break down'" (Sebeok, 1988, p. 65; see also note 1 in Sebeok, 1988, p. 72). "According to this, in general, a fungus/interpretant is mediately determined by an animal/sign, which is determined by a plant/object (but plant/fungus are likewise variant life forms, of course, just as object/interpretant are both sign variants)" (Sebeok, 1999b, p. 391).

In the framework of endosemiotics, a special area of *immunosemiotics* (and semioimmunology) has also been noted as a field dealing with the immunological code, immunological memory and recognition (Sebeok, 1997a, p. 438, 2001c, p. 21; Sercarz et al., 1988).

As E. Baer (1987, p. 206) says, "Sebeok's work marks a transition of semiotics from a one-sided subjection to the linguistic model to a biologically oriented investigation of Umwelt." In his papers of different topics, Sebeok has tried to emphasize and demonstrate the existence of semiosic phenomena in non-human organisms, and to analyse the biological basis of various sign processes. This includes, among others, the biological derivation of non-verbal art forms (Sebeok, 1984).

While discussing the view held by semiotics of culture that the appearance of culture provides the semiotic threshold, there is surprisingly much in what Sebeok incorporates from the Tartu School. Particularly, the concept of modelling systems as introduced into semiotics by Juri Lotman and his colleagues (e.g., the Kääriku Summer Schools on Secondary Modelling Systems, in 1960s). The book *Forms of Meaning* (Sebeok & Danesi, p. 2000) uses the concept of the modelling system as the central one. Also, "many of Sebeok's studies constitute fundamental continuations of Uexküll's project of Umwelt research" (Baer, 1987, p. 205).

When describing the semiotic behaviour of animals and other organisms, Sebeok does not apply a gradualistic approach. He sharply distinguishes life as the arena of semioses from non-life, as well as human semioses from non-human semiosis.

In addition to specifically biosemiotic problems, Sebeok also touches, in some of his writings, on the area of representations of (and approaches to) nature in cultures (as can be illustrated, for instance, by the quotation that heads this paper). This field, nowadays known as (cultural) *ecosemiotics,* should be taken as different from biosemiotics, because it does not deal with biological problems and belongs rather to the domain of the semiotics of culture.

The core statements of biosemiotics

It will be fascinating to try to formulate briefly, in a thesis-like form, the main statements of Sebeok on biological semiotics.[14] The version of these 'theses on biosemiotics' that follows below is compiled from his various writings on the issue. Among his own papers, the article 'Signs, bridges, origins' includes some of these statements, formulated in terms of 'theorems' and 'lemmas' (Sebeok, 1996b, also published in a slightly edited version in Sebeok, 2001c, pp. 59–73).

(1) *Life is semiosis.* Semiosis, or a triadic cooperative production involving a sign, its object, and its interpretant , is as much a criterial attribute of all life as is the ability to metabolize (Umiker-Sebeok & Sebeok, 1980, p. 1).
(2) *Umwelt is a model.* The recalcitrant term Umwelt had best be rendered in English by the word model (Sebeok, 1988, p. 72). All, and only, living entities incorporate a species-specific model (Umwelt) of their universe (Sebeok 1996b: 102).
(3)) *There exists a global communicative network in the biosphere, formed in its lowest level by bacteria.* The earliest, smallest known biospheric module with semiosic potential is a single bacterial cell. The largest, most complex living entity may be Gaia. Both units at the polar ends display general properties of autopoietic entities, but it is now bacteria that merit, in my opinion, special consideration on the part of all who would work at semiotics professionally (Sebeok, 2001c, p. 12).
(4) *Protists, plants, fungi, and animals represent different basic communication strategies, and accordingly, correspondent branches of biosemiotics are relevant.* Just as there are different sorts of strategies for metabolic activity, there are also various kinds of communication devices (Umiker-Sebeok & Sebeok, 1980, p. 1).

---

14. Two other recent attempts to formulate the main theses of biosemiotics (mainly referring to J. Hoffmeyer's writings) can be found in Emmeche et al., 2002, pp. 13-24, and Stjernfelt, 2002.

(5) *Endosemiosis occurs in organism — with multiple (genetic, immune, metabolic, neural) codes.* These four codes (with references to relevant literature) are mentioned, e.g., in Sebeok (1996b, pp. 107—108).
(6) *Symbiosis is a token of semiosis.* The biologist s notion of symbiosis is equivalent to the philosopher s notion of semiosis (Sebeok, 1988, p. 72). Inasmuch as processes of sign transmission outside and inside organisms are at play, it appears not unreasonable to suppose symbiosis to be a token of semiosis and endosymbiosis to be a token of endosemiosis (Sebeok, 1996b, p. 102).
(7) *Language appears with syntax. There are no syntactic structures in animal sign systems.* What we know of zoosemiotic processes furnishes no evidence of syntactic structures, not even in any of the alloprimates (Sebeok, 1996b, p. 108).

These and a couple of other statements of the same kind form some important knots in the network of Sebeok's ideas, which are also illustrated by him through a large number of examples, references and citations from a big variety of sources he has used in compiling his texts.

Building a field: The biosemiotic web

Despite the many fields to which Sebeok contributed, his work in biosemiotics was evidently seen by him to be of central importance. When he understood that the building of semiotic biology would mean a paradigmatic change, he consciously wanted to establish the necessary attributes for this area to become a recognised independent field of research. This means, above all, the publications, particularly thematic volumes and monographs, and the history of the field.

An important event in this direction has been the publication of English translations of Jakob von Uexküll's two books as special issues of *Semiotica* (vol. 42(1), 1982, & vol. 89(4), 1992). Certainly Sebeok's role has been most helpful in getting Uexküll acknowledged as one of the major classics of contemporary semiotics.

Sebeok was an engine behind the first specialised meetings on biosemiotics, in 1991 and 1992. These took place in Glottertal, a village near Freiburg am Main in Germany.[15] As Jesper Hoffmeyer (2002, p. 385) has said, "these early Glotterbad meetings were perhaps especially important because they left an impression on everybody that biosemiotics was now for real."[16]

---

15. About that meeting, see also in Hoffmeyer, 2002, pp. 384-385, and Sebeok, 2001a, p. 65; 2001c, p. 170.
16. Glottertal meeting in 1992 was also the one were I first met Thomas Sebeok. After that we had a chance to meet quite many times — in Tartu (Sebeok, 1997, 1999), in Imatra (Sebeok, 1998, 1999, 2000), in Toronto (Sebeok, 1997), in Siena (Sebeok, 1998), in Dresden (Sebeok, 2000), in Bloomington (Sebeok, 1999). I was particularly glad to make a two-week visit to Bloomington in 1999, where I spent many hours every day at the Sebeoks' home and could work through many tens of meters of Tom's bookshelves (Fig. 3).

In 1992, the first collection of papers on biosemiotics has been published under his editorship (together with his wife Jean Umiker-Sebeok) (Sebeok & Umiker-Sebeok, 1992).

Sebeok's support of biosemiotic publications has been remarkable. For instance, this concerns a series of writings by Thure von Uexküll, a translation of Giorgio Prodi's work, the spread of phytosemiotic papers by Martin Krampen (1981, and its several later versions), and the English translation of a book by Jesper Hoffmeyer (1996). On the latter, Sebeok organised a series of reviews that were published as a special issue of *Semiotica* (vol. 120(3/4), 1998). Without Sebeok's enthusiastic support, the two large special volumes on biosemiotics would not have been published – *Semiotica* vol. 127(1/4), 1999, edited by J. Hoffmeyer and C. Emmeche, and vol. 134(1/4), edited by K. Kull.

Sebeok devoted many of his conference lectures to the various aspects of history of biosemiotics. Large number of his writings include descriptions of the work and views of Jakob von Uexküll (e.g., Sebeok, 1977, 1998). In several of his papers he tried to frame the history of biosemiotics in general (Sebeok, 1996a, 1999a, 2001a).[17] During the second half of 1990s, a conscious attempt has been made to produce a systematic series of papers on the history of biosemiotics. This has resulted in a series of papers that reviewed the biological aspects in the works of semiotics classics — Peirce, Morris, Jakobson, Lotman, and few others.[18] We talked about this plan several times, during our meetings in Imatra and elsewhere.

Still, despite the large number of writings, there seem to be a couple of biosemiotic problems that Sebeok almost did not touch. One of these concerns his avoidance of the topic of (biological) epistemology, otherwise quite intensively discussed in biosemiotic literature (e.g., Hoffmeyer, 1996; Pattee, 2001; Vehkavaara, 2002). I would hypothesize that Sebeok's position has to do with his use of the concepts of model and modelling. Indeed, "in a biosemiotic paradigm, the function of singularized modelling is viewed as a general strategy for giving the perception of single objects, unitary events, individual feelings, etc. a knowable form …. Signs are … 'recognition-enhancing forms,' which allow for the detection of relevant incoming sensory information in a patterned fashion" (Sebeok & Danesi, 2000, p. 20). Also, a very interesting paper "What do we know about signifying behavior in the domestic cat (*Felis catus*)?" (Sebeok, 2001c, pp. 74–96) asks and sheds light on several questions about the ways of knowing the worlds of other organisms (Fig. on the cover).

Another problem, scarcely analysed by Sebeok, is the methodology of biosemiotic inquiry. One can be referred to the works of Jakob von Uexküll as the ones that provide the necessary approach. However, in addition to the points described by Sebeok, the practical questions of how the non-verbal sign systems of non-human organisms should be studied, and which are the criteria that allow us to assign them

---

17. However, he never wrote a general review on the history of biosemiotics (cf. Kull, 1999a).
18. See footnote 3.

the usage of meaning, still require a profound elaboration. Otherwise the step from ethology to biosemiotics is hardly thinkable.

Building biosemiotics exceeds the borders of biology. "Indeed, there is a lot of work to do for serious philosophy, considering how many central philosophical topics — of mind, language, epistemology, and metaphysics — that cannot remain unaffected by the biosemiotic turn" (Emmeche, 2002, p.158).

Credo

It is interesting to remember how Sebeok has characterised himself:

> I firmly believe that there are, and should be, two complementary sorts of scholars: I call them moles and bees. Moles have tough nuzzles and powerful forefeet for burrowing ever deeper in one and the same spot. Such a profound scholarly mole I am not.
>
> Bees, by contrast, dart solitary from flower to flower, sipping nectar, gathering pollen from flowers, serendipitously fertilizing whatever they touch. I fancy that I have always been something of a, maybe superficial, academic *Apis mellifera*. This honeybee is the semiotic species par excellence, possessed, next to our own, of the most elaborate social communication system thus far recognized by ethologists. Too, it seldom stings unless its budget is threatened. (Sebeok, 1995, p. 121)

There also exists an earlier version of this characterisation that uses an example of laboratory rats.[19] It is important to understand that there is much more than an allegory in these slightly humoristic accounts. Because, according to Sebeok, the life process is the same in all living, and since this is a semiotic process, these comparisons state something about the ways of life in general. This can be illustrated by a reference in his book entitled *I Think I Am a Verb* to his two daughters as his "immediate and emotional *interpretants*" (Sebeok, 1986a, p. vii; my emphasis).

Thomas Sebeok's credo is something that we should all learn from him. In his own words (Sebeok, 1995, p.125; my emphasis):

> To conclude ... on a semiotic note, and drawing on an image from Samuel Butler, I would observe that an academic is a sign's way of spawning further, more developed academics. The administration's task is to ensure that this process works smoothly. There are two fundamental strategies to accomplish these ends: first, by *publishing and teaching as much as possible;* and, equally important, by *doing one's best to facilitate the success of one's colleagues* in these respects. These are the only things I have ever wanted to do in my academic life.

---

19. "There appear to be two antipodal sorts of bookmen. There are those who derive endless delight from their solitary pleasure, which they pursue like self-stimulating laboratory rats, with electrodes implanted in their anterior hypothalamus, unceasingly bar-pressing in preference to any other activity. Then there are those of us whose bar-pressing habit is rewarded solely by a change in the level of illumination – in a word, novelty" (Sebeok, 1986a, p. x).

References

Anderson, M. (1990). Biology and semiotics. In: W. A. Koch (Ed.). *Semiotics in the Individual Sciences*, Part I.(pp. 254–281). Bochum: Universitätsverlag Dr. N. Brockmeyer.
Anderson, M. (2000). Sharing G. Evelyn Hutchinson's fabricational noise. *Sign Systems Studies, 28,* 388–396.
Anderson, M., Deely, J., Krampen, M., Ransdell, J., Sebeok, T. A., & Uexküll, T. von. (1984). A semiotic perspective on the sciences: steps toward a new paradigm. *Semiotica, 52,* 7–47.
Baer, E. (1987). Thomas A. Sebeok's doctrine of signs. In: M. Krampen, K. Oehler, R. Posner, T.A. Sebeok, & T. von Uexküll (Eds.), *Classics of Semiotics* (pp.181–210). New York: Plenum Press.
Bernard, J. (2002). In memory of Thomas A: Sebeok (1920–2001). *Zeitschrift für Semiotik, 24*(1).
Bernard, J., Deely, J., Voigt, V., & Withalm, G. (Eds.). (1993). *Symbolicity: Proceedings of the International Semioticians' Conference in Honor of Thomas A. Sebeok's 70th Birthday.* Lanham: University Press of America.
Bonfante, G., Sebeok, T. A. (1944). Linguistics and the age and area hypothesis. *American Anthropologist, 46,* 382–386.
Bouissac, P., Herzfeld, M., & Posner, R. (Eds.). (1986). *Iconicity: Essays on the Nature of Culture: Festschrift für T. A. Sebeok zum 65. Geburtstag.* Tübingen: Staufenburg Verlag.
Brauckmann, S. (2000). Steps towards an ecology of cognition: A holistic essay. *Sign Systems Studies, 28,* 397–420.
Carmeli, Y. S. (2002). On the 'culture' dimension in a biosemiotic inquiry. *Semiotica, 141*(1/4), 415–430.
Cimatti, F. (2000). The circular semiosis of Giorgio Prodi. *Sign Systems Studies, 28,* 351–379.
Danesi, M. (1990). Zoosemiotics: Thomas Sebeok fashions a field of scientific inquiry. In T. Sebeok, *Essays in Zoosemiotics* (pp. 7-9). Toronto: Toronto Semiotic Circle.
Danesi, M.(2000). The biosemiotic paradigm of Thomas A. Sebeok. In E.Tarasti (Ed.), *Commentationes in honorem Thomas A. Sebeok octogenarii A. D. MM editae* (pp. 5–29). Imatra: International Semiotics Institute.
Danesi, M. (2001). Hediger through Sebeok: An introduction to the biosemiotic paradigm. In T. Sebeok *The Swiss Pioneer in Nonverbal Communication Studies: Heini Hediger (1908–1992)* (pp. 7–13). New York: Legas.
Deely, J. (1995a). Quondam magician, possible martian, semiotician: Thomas Albert Sebeok. In N. Tasca (Ed.), *Ensaios em homenagem a: Essays in Honor of Thomas A. Sebeok* (pp. 17-26). Porto: Almeida.
Deely, J. (Ed.). (1995b). *Thomas A. Sebeok: Bibliography 1942–1995.* Bloomington: Eurolingua.
Deely, J. (1998). Sebeok, Thomas A. In P. Bouissac (Ed.), *Encyclopedia of Semiotics* (pp. 557–559). New York: Oxford University Press.
Emmeche, C. (2002). Taking the semiotic turn, or how significant philosophy of biology should be done. *Sats – Nordic Journal of Philosophy, 3*(1), 155–162.
Emmeche, C., Kull, K., & Stjernfelt, F. (2002). *Reading Hoffmeyer, Rethinking Biology.* Tartu: Tartu University Press.
Hoffmeyer, J. (1996). *Signs of Meaning in the Universe.* Bloomington: Indiana University Press.
Hoffmeyer, J. (2002). Obituary: Thomas A. Sebeok. *Sign Systems Studies, 30*(1), 383–386.
Hoffmeyer, J., & Emmeche, C. (1991). Code-duality and the semiotics of nature. In M. Anderson & F. Merrell (Eds.), *On Semiotic Modeling* (pp. 117–166). Berlin: Mouton de Gruyter.
Krampen, M. (1981). Phytosemiotics. *Semiotica, 36*(3/4), 187–209.
Kalmus, H. (1950). A cybernetic aspect of genetics. *Journal of Heredity, 41,* 19–22.
Kalmus, H. (1962). Analogies of language to life. *Language and Speech, 5*(1), 15–25.
Kull, K. (1999a). Biosemiotics in the twentieth century: A view from biology. *Semiotica, 127*(1/4), 385–414.
Kull, K. (1999b). Towards biosemiotics with Yuri Lotman. *Semiotica, 127*(1/4), 115–131.
Kull, K. (1999c). On the history of joining bio with semio: F. S. Rothschild and the biosemiotic rules. *Sign Systems Studies, 27,* 128–138.
Kull, K. (2001). Jakob von Uexküll: An introduction. *Semiotica, 134*(1/4), 1–59.
Kull, K., Lotman, M., & Torop, P. (2002). In memoriam Thomas Albert Sebeok. *Trames, 6*(1), 110–112.
Moriarty, S. (1994). Visual communication as a primary system. *Journal of Visual Literacy, 14*(2), 11–21.
Nöth, W. (2000). *Handbuch der Semiotik* (2te Auflage). Stuttgart: J. B. Metzler.
Nuessel, F. (2000). A sign is just a sign: Sebeok revisited in Italian translation. *Semiotica, 131*(3/4), 201–216.
O'Connor, A. (2002). Thomas Sebeok, 81, debunker of ape-human speech theory, is dead. *The New York Times,* January 2.
Pattee, H. H. (2001). Irreducible and complementary semiotic forms. *Semiotica, 134*(1/4), 341–358.
Petrilli, S. (1999a). The biological basis of Victoria Welby's significs. *Semiotica, 127*(1/4), 23–66.
Petrilli, S. (1999b). Charles Morris's biosemiotics. *Semiotica, 127*(1/4), 67–102.
Petrilli, S. (2002). Una vita per i segni della vita: Thomas A. Sebeok. *Athanor, 5,* 19–20.
Petrilli, S., & Ponzio, A. (2001). *Thomas Sebeok and the Signs of Life.* Cambridge: Icon Books.
Ponzio, A., & Petrilli, S. (2002). *I segni e la vita: La semiotica globale di Thomas A. Sebeok.* Milano: Spirali.
Ruse, M. (1998). Zoosemiotics. In P. Bouissac (Ed.), *Encyclopedia of Semiotics* (p. 652). New York: Oxford University Press,.
Santaella Braga, L. (1999). Peirce and biology. *Semiotica, 127*(1/4), 5–21.
Sebeok, T. A. (1962). Coding in the evolution of signalling behavior. *Behavioral Science, 7,* 430–442.
Sebeok, T. A. (1963). Communication among social bees; porpoises and sonar; man and dolphin. *Language, 39,* 448–466.
Sebeok, T. A. (1965a). Animal communication. *Science, 147,* 1006–1014.

Sebeok, T. A. (1965b). Zoosemiotics: A new key to linguistics. *The Review, 7,* 27–33.
Sebeok, T. A. (1965c). Coding in animals and man. *ETC, 22,* 330–249.
Sebeok, T. A. (Ed.). (1968). *Animal Communication: Techniques of Study and Results of Research.* Bloomington: Indiana University Press.
Sebeok, T. A. (1969). Semiotics and ethology. In T.A. Sebeok & A. Ramsay (Eds.), *Approaches to Animal Communication* (pp. 200-231). The Hague: Mouton.
Sebeok, T. A. (1972). *Perspectives in Zoosemiotics.* The Hague: Mouton.
Sebeok, T. A. (1976). *Contributions to the Doctrine of Signs.* Bloomington: Indiana University Press.
Sebeok, T. A. (1977). Ecumenicalism in semiotics. In T. A. Sebeok (Ed.), *A Perfusion of Signs* (pp.180–206). Bloomington: Indiana University Press.
Sebeok, T. A. (1980). Looking in the destination for what should have been sought in the source. In T. A. Sebeok & J. Umiker-Sebeok (Eds.), *Speaking of Apes: A Critical Anthology of Two-Way Communication with Man* (pp. 407–427). New York: Plenum Press.
Sebeok, T. A. (1984). Prefigurements of art. In: J. Pelc, T. A. Sebeok, E. Stankiewicz, & T. G. Winner (Eds.), *Sign, System and Function: Papers of the First and Second Polish-American Semiotics Colloquia* (pp. 361–362). Berlin: Mouton Publishers,
Sebeok, T. A. (1986a). *I Think I Am a Verb: More Contributions to the Doctrine of Signs.* New York: Plenum Press.
Sebeok, T. A. (Ed.). (1986b). *Encyclopedic Dictionary of Semiotics,* Vols. 1–3. Berlin: Mouton de Gruyter.
Sebeok, T. A. (1987). Karl Bühler. In: M. Krampen, K. Oehler, R. Posner, T. A. Sebeok, & T. von Uexküll (Eds.), *Classics of semiotics* (pp. 129–145). New York: Plenum.
Sebeok, T. A. (1988). 'Animal' in biological and semiotic perspective. In T. Ingold (Ed.), *What is an Animal?* (pp. 63–76). London: Unwin Hyman.
Sebeok, T. A. (1990). *Essays in Zoosemiotics.* Toronto: Toronto Semiotic Circle.
Sebeok, T. A. (1991). *A Sign is Just a Sign.* Bloomington: Indiana University Press.
Sebeok, T. A. (1992). 'Tell me, where is fancy bred?': The biosemiotic self. In T. A. Sebeok, & J. Umiker-Sebeok (Eds.) *Biosemiotics: The Semiotic Web 1991*(pp. 333–343). Berlin: Mouton de Gruyter.
Sebeok, T. A. (1994). *Signs: An Introduction to Semiotics.* Toronto: University of Toronto Press.
Sebeok, T. A. (1995). Into the rose-garden. In J. Deely (Ed.), *Thomas A. Sebeok: Bibliography 1942–1995* (pp.116–125). Bloomington: Eurolingua.
Sebeok, T. A. (1996a). Galen in medical semiotics. *Interdisciplinary Journal for Germanic Linguistics and Semiotic Analysis, 1*(1): 89–111.
Sebeok, T. A. (1996b). Signs, bridges, origins. In J. Trabant (Ed.), *Origins of Language* (pp. 89–115). Budapest: Collegium Budapest.
Sebeok, T. A. (1997a). The evolution of semiosis. In: R. Posner, K. Robering, & T. A. Sebeok (Eds.), *Semiotics: A Handbook on the Sign-Theoretic Foundations of Nature and Culture,* Vol. 1 (pp. 436–446). Berlin: Walter de Gruyter.
Sebeok, T. A. (1997b.) Give me another horse. In R. Capozzi (Ed.), *Reading Eco: An Anthology* (pp. 276–282). Bloomington: Indiana University Press.
Sebeok, T. A. (1998). The Estonian connection. *Sign Systems Studies, 26,* 20–41.
Sebeok, T. A. (1999a). Editor's note: Towards a prehistory of biosemiotics. *Semiotica, 127*(1/4), 1–3.
Sebeok, T. A. (1999b). The life science and the sign science. *Applied Semiotics, 6/7,* 386–393.
Sebeok, T. A. (2000). The music of spheres. *Semiotica, 128*(3/4), 527–535.
Sebeok, T. A. (2001a). Biosemiotics: Its roots, proliferation, and prospects. *Semiotica, 134*(1/4), 61–78.
Sebeok, T. A. (2001b). *The Swiss Pioneer in Nonverbal Communication Studies: Heini Hediger (1908–1992).* New York: Legas.
Sebeok, T. A. (2001c). *Global Semiotics.* Bloomington: Indiana University Press.
Sebeok, T. A. (2001d). Biosemiotics. In P. Cobley (Ed.), *The Routledge Companion to Semiotics and Linguistics* (pp. 163–164). London: Routledge.
Sebeok, T. A. (2002). La semiosfera come biosfera. *Athanor, 5:* 11–18.
Sebeok, T. A., & Danesi, M. (2000). *The Forms of Meaning: Modeling Systems Theory and Semiotic Analysis.* Berlin: Mouton de Gruyter.
Sebeok, T. A., & Ramsay, A. (Eds.). (1969). *Approaches to Animal Communication.* The Hague: Mouton.
Sebeok, T. A., & Rosenthal, R. (Eds.). (1981). The Clever Hans Phenomenon: Communication with Horses, Whales, Apes, and People. *Annals of the New York Academy of Sciences, 364.* New York: New York Academy of Sciences.
Sebeok, T. A., & Umiker-Sebeok, J. (Eds.). (1980). *Speaking of Apes: A Critical Anthology of Two-Way Communication with Man.* New York: Plenum Press.
Sebeok, T. A., & Umiker-Sebeok, J. (Eds.). (1992). *Biosemiotics: The Semiotic Web 1991.* Berlin: Mouton de Gruyter.
Sercarz, E. E., Celada, F., Mitchison, N. A., & Tada, T. (Eds.). (1988). *The Semiotics of Cellular Communication in the Immune System.* Berlin: Springer.
Shintani, L. (1999). Roman Jakobson and biology: 'A system of systems'. *Semiotica, 127*(1/4), 103–113.
Smith, W. J. (1974). Zoosemiotics: Ethology and the theory of signs. In T. A. Sebeok (Ed.), *Current Trends in Linguistics,* 12 (pp.561–626). The Hague: Mouton.
Stepanov, Y. S. (1971). *Semiotika.* Moskva: Nauka.
Stjernfelt, F. (2002). Tractatus Hoffmeyerensis: Biosemiotics as expressed in 22 basic hypotheses. *Sign Systems Studies, 30*(1), 337–345.

Tarasti, E. (Ed.). (2000). *Commentationes in honorem Thomas A. Sebeok octogenarii A. D. MM editae*. Imatra: International Semiotics Institute.
Tasca, N. (Ed.). (1995). *Ensaios em homenagem a: Essays in Honor of Thomas A. Sebeok*. Porto: Almeida.
Turovski, A. (2000). The semiotics of animal freedom: A zoologist's attempt to perceive the semiotic aim of H. Hediger. *Sign Systems Studies, 28*: 380–387.
Umiker-Sebeok, J., & Sebeok, T. A. (1980). Introduction: Questioning apes. In T. A. Sebeok, & J. Umiker-Sebeok (Eds.) *Speaking of Apes: A Critical Anthology of Two-Way Communication with Man* (pp.1–59). New York: Plenum Press.
Vehkavaara, T. (2002). Why and how to naturalize semiotic concepts for biosemiotics. *Sign Systems Studies, 30*(1), 293–313.
Willis, J. C. (1922). *Age and Area: A Study in Geographical Distribution and Origin of Species*. Cambridge: Cambridge University Press.

Sebeok's Semiosic Universe and Global Semiotics

Susan Petrilli[1]

In the belief that the universe is a semiosic web Sebeok placed no limits on semiotic inquiry as he chose to conduct it. In terms of fields of semiosis and objects of analysis Sebeok's interests ranged from his studies in biology to his reflections relative to linguistics, cybernetics, artificial intelligence, and musements relative to narrative aesthetic signs. He was interested in ecology and the signs of deception, sometimes inventing new disciplines to focus on specific aspects in the life of signs, such as "zoösemiotics" and "biosemiotics." Even more significant is his method of inquiry and its ontological foundations. Sebeok was a convinced critic of code semiotics and the restriction of its focus to the human social world, privileging instead the instruments of interpretation semiotics which best accounts for his axiom that life is the criterial attribute of semiosis. Working in such a direction he created the conditions for his "global semiotics," an approach which favours conceiving the totality of semiosis, the great semiobiosphere, in a detotalizing perspective, that is, in its interrelated plurality and diversification. The play of musement is a specifically human capacity theorized by Sebeok according to his biosemiotic interpretation of the concept of language as a syntactic modeling device. And let us add that he exemplifies the play of musement in his life as a researcher and critical interpreter of signs beyond prejudicial boundaries, even perhaps beyond the signs of life.

1. Sebeok's semiosic universe

1.1. Signs, other signs and more...
Thomas A. Sebeok began his studies in higher education during the second half of the thirties at Cambridge. He was particularly influenced by *The Meaning of Meaning* (1923) by Charles K. Ogden and Ivor A. Richards long before it became a classic in semiotics. Also he can boast of having benefitted from direct contact with two great masters of the sign who were also his teachers in different ways: Charles Morris and Roman Jakobson (see chpt. 5, "Vital Signs," in Sebeok, 1986, and the parts dedicated to these figures in Sebeok, 1979b, 1985, 1991c, 1992).

The variegated aspects and parts of the manifold "semiosic universe" as they emerge from Sebeok's semiotic research include:

---

1. Susan Petrilli is Associate Professor of Semiotics at the Dipartimento di Pratiche linguistiche e analisi di testi, University of Bari. Mailing address: Dipartimento di Pratiche linguistiche e analisi di testi, Facoltà di Lingue e Letterature Straniere, Università di Bari, Via Garruba n. 6, Bari-70100, Italy. Home page: www.lingue.uniba.it/dip-plat/indice; Email: s.petrilli@lingue.uniba.it.

— The life of signs and the signs of life as they appear today in the biological sciences: the signs of animal life and of specifically human life, of adult life, and of the organism's relations with the environment, and the signs of normal and pathological forms of dissolution and deterioration of communicative capabilities.

— Human verbal and nonverbal signs. Human nonverbal signs include signs which depend on natural languages and signs which, on the contrary, are not dependent on natural language and which, therefore, cannot be accounted for by the categories of linguistics. These include the signs of "parasitic" languages, such as artificial languages, the signs of "gestural languages," such as the sign language of Amerindian and Australian aborigines, and the language of deaf-mutes; the signs of infants, and the signs of the human body both in its more culturally dependent manifestations as well in its natural-biological manifestations.

— Human intentional signs controlled by the will, and unintentional, unconscious signs such as those that pass in communication between human beings and animals in "Clever Hans" cases (cf. Sebeok, 1979, 1986). Here, animals seem capable of certain performances (e.g., counting), simply because they respond to unintentional and involuntary suggestions from their trainers. This group includes signs at all levels of conscious and unconscious life, and signs in all forms of lying (which Sebeok identifies and studies in animals as well), deceit, self-deceit, and good faith.

— Signs at a maximum degree of plurivocality and, on conversely, signs that are characterized by univocality and which, therefore, are signals.

— Signs viewed in all their shadings of indexicality, iconicity, and symbolicity.

— Finally, "signs of the masters of signs." Those through which it is possible to trace the origins of semiotics (for example, in its ancient relation to divination and to medicine), or through which we may identify the scholars who have contributed directly or indirectly (as "cryptosemioticians") to the characterization and development of this science, or "signs of the masters of signs" through which we may establish the origins and development of semiotics relatively to a given nation or culture, as in Sebeok's study on semiotics in the United States. "Signs of the masters of signs" also includes the narrative signs of anecdotes, testimonies and personal memoirs that reveal these masters not only as scholars but also as persons — their character, behavior, everyday habits. Not even these signs, "human, too human," escape Sebeok's semiotic interests.

All this is a far cry from the limited science of signs as conceived in the Saussurean tradition!

1.2. Critique of the pars pro toto error

As a fact of signification the entire universe enters what Sebeok (2001)dubbed "global semiotics."[2] In such a global perspective semiotics is the place where the "life

---

2. The expression "Global Semiotics" is the title of a plenary lecture delivered by Sebeok on June 18, 1994 as Honorary President of the Fifth Congress of the International Association for Semiotic Studies, held at the University of California, Berkeley, now in Sebeok, Global Semiotics, 2001.

sciences" and the "sign sciences" converge. This means that *signs* and *life* converge. Therefore global semiotics is the place where human consciousness fully realizes that the human being is a sign in a universe of signs.

Sebeok extends the traditional boundaries of sign studies or semiology, providing an approach to the study of signs that is more comprehensive than possible in semiology. The limit of *semiology*, the science of the signs as projected by Saussure, consists of the fact that it is based on the verbal paradigm and is vitiated by the mistake of *pars pro toto* — in other words, Saussurean semiology mistakes the part (human signs and in particular verbal signs) for the whole (all possible signs, human and nonhuman). On the basis of such a mystification, semiology incorrectly claims to be the general science of signs. On the contrary, when the general science of signs chooses the term "semiotics" for itself, the aim is to take its distances from semiology and its errors. Sebeok dubs the semiological tradition in the study of signs the "minor tradition," and promotes what he dubs as the "major tradition" represented by John Locke and Charles S. Peirce, as well as by the ancients, Hippocrates and Galen and their early studies on signs and symptoms. Semiotics, therefore, is at once recent when considering the determination of its status and awareness of its wide-ranging possible applications, and ancient when considering that its roots can be traced back at least, following Sebeok (1979), to the theory and practice of Hippocrates and Galen.

Through his numerous publications Sebeok has propounded a wide-ranging vision of semiotics that coincides with the study of the evolution of life. After Sebeok's work (which is largely inspired by Charles S. Peirce, Charles Morris and Roman Jakobson), both the conceptions of the semiotic field and of the history of semiotics have changed. Thanks to Sebeok, semiotics at the beginning of the new millennium has extended its horizons to be far broader than envisaged during the first half of the 1960s.

Sebeok's approach to the *life of signs* is global or holistic and may be immediately associated with his concern for the *signs of life*. In his view *semiosis* and *life* coincide. That semiosis originates with the first stirrings of life, leads to the formulation of an axiom which he believes is cardinal to semiotics: semiosis is the criterial attribute of life.

"Global semiotics" (Sebeok, 2001) provides a meeting point and an observation post for studies on the life of signs and the signs of life. In line with the "major tradition" in semiotics, Sebeok's global approach to sign life presupposes his critique of anthropocentric and glottocentric semiotic theory and practice. In his explorations of the boundaries and margins of the science or *doctrine* of signs (as he also calls it), Sebeok opens the field to include *zoösemiotics* (a term he introduced in 1963), or, even more broadly *biosemiotics*, on the one hand, and endosemiotics (semiotics of sign systems such as the immunological, or the neuronal, cf. Thure von Uexküll, "Endosemiosis," in Posner, Robering, & Sebeok, 1997, Vol. 1, pp. 464-487) on the other. In Sebeok's conception, the sign science is not only the "science qui étude la vie des signes au sein de la vie sociale" (Saussure, 1916, pp. 26), that is, the study of communication in culture, but also the study of communicative behavior in a biosemiotic perspective. Consequently, Sebeok's global semiotics is characterized by a maximum broadening of competencies.

1.3. Crossing over semiosic boundaries

Sebeok's article "The Evolution of Semiosis" (Sebeok, 1997-98; and in Sebeok, 1991) opens with the question "what is semiosis?" and the answer begins with a citation from Peirce. Sebeok observes that Peirce's description (*CP* 5.473) of semiosis or the "action of a sign" as an irreducibly triadic process or relation (sign, object, and interpretant), focuses particularly upon how the interpretant is produced, therefore it concerns that which is involved in understanding or in the teleonomic (that is, goal-directed) interpretation of the sign.

Not only is there a sign which is a sign of something else, but also a *somebody* a "*Quasi-interpreter*" (*CP* 4.551) which takes something as a sign of something else. Peirce analyzed the implications of this description further when he said that: "It is of the nature of a sign, and in particular of a sign which is rendered significant by a character which lies in the fact that it will be interpreted as a sign. Of course, nothing is a sign unless it is interpreted as a sign" (*CP* 2.308). And again: "A sign is only a sign in actu by virtue of its receiving an interpretation, that is, by virtue of its determining another sign of the same object" (*CP* 5.569).

From the viewpoint of the interpretant and, therefore, of sign-interpreting activity or process of inferring from signs, *semiosis* may be described in terms of *interpretation*. Peirce specifies that all "signs require at least two *Quasi-minds*; a *Quasi-utterer* and a *Quasi-interpreter*" (*CP* 4.551). The interpreter, mind or quasi-mind, "is also a sign" (Sebeok, 2001b, p. 14), exactly a response, in other words, an interpretant: an interpreter is a responsive "somebody."

In his article, "The Evolution of Semiosis," Sebeok continues his answer to the question "what is semiosis?" with a citation from Morris (1946, 1949, 1971) who defined semiosis as "a process in which something is a sign to some organism." This definition implies effectively and ineluctably, says Sebeok, the presence of a living entity in semiosic processes. And this means that semiosis appeared with the evolution of life.

> For this reason one must, for example, assume that the report, in the King James version of the Bible (Genesis I.3), quoting God as having said "Let there be light," must be a misrepresentation; what God probably said was "let there be photons," because the sensation of perception of electromagnetic radiation in the form of optical signals (Hailman,1997, pp. 56-58), that is, luminance, requires a living interpreter, and the animation of matter did not come to pass much earlier than about 3,900 million years ago. (Sebeok, 1997-98, p. 436)

In Morris's view the living entity implied in semiosis is a macro-organism; according to Sebeok's global semiotics instead it may be a cell, a portion of a cell, or a genoma.

In "The Evolution of Semiosis" Sebeok examines the question of the cosmos before semiosis and after the beginning of the Universe and refers to the regnant paradigm of modern cosmology, i. e., the Big Bang theory. Before the appearance of life on our planet — the first traces of which date back to the so-called Archaean Aeon, from 3,900 to 2,500 million years ago — there were only physical phenomena involving interactions of non-biological atoms, later of inorganic molecules. Such interactions may be described as *quasi-semiotic*. But the notion of *quasi-semiosis* must be distinguished from *protosemiosis* as understood by the Italian oncologist Giorgio Prodi[3] (cf. Prodi, 1977; the milestone volume *Biosemiotics*, edited by Sebeok and Umiker-Sebeok, 1992, is dedicated to Prodi who is described as a "bold trailblazer of contemporary biosemiotics"). In fact, in the case of physical phenomena, the notion of protosemiosis is metaphorical. In Sebeok's view, semiosis concerns life. He distinguishes between nonbiological interactions and "primitive communication" which refers to the transfer of information with endoparticles, as in neuron assemblies where in modern cells transfer is managed by protein particles.

Since there is not a single example of life outside our terrestrial biosphere, the question of whether there is life/semiosis elsewhere in our galaxy, let alone in deep space, is wide open. Therefore, says Sebeok, one cannot but hold "exobiology semiotics" and "extraterrestrial semiotics" to be twin sciences that so far remain without a subject matter.

In the light of present-day information, all this implies that at least one link in the semiosic loop must necessarily be a living and terrestrial entity: this may even be a mere portion of an organism or an artifact extension fabricated by human beings. After all, semiosis is terrestrial biosemiosis. A pivotal concept in Sebeok's research, as we have already stated, is that semiosis and life coincide. Semiosis is considered the criterial feature that distinguishes the animate from the inanimate, and sign processes have not always existed in the course of the development of the universe: sign processes and the animate originated together with the development of life.

Identification of semiosis and life invests semiotics with a completely different role from that conceived by Eco (1975) when he described the conjunction between semiosis and life as concerning "the inferior threshold of semiotics." In Eco's view semiotics is a cultural science. Sebeok interprets and practices semiotics as a life science, as biosemiotics; biosemiotics cannot be reduced to its interpretation as a mere "sector" of semiotics.

1.4. Global semiotics

For Sebeok, semiotics is more than just a science that studies signs in the sphere of socio-cultural life, in other words, "la science qui étude la vie des signes au sein de la vie sociale" (Saussure, 1916 p. 26). Before contemplating the signs of unintentional communication (semiology of signification), semiotics was further limited by an

---

3. Giorgio Prodi (1928-1987) "was, on the one hand, one of his country's leading medical biologists in oncology, while he was, on the other, a highly original contributor to semiotics and epistemology, the philosophy of language and formal logic, plus a noteworthy literary figure. Prodi's earliest contribution to this area [immunosemiotics, an important branch of biosemiotics], [is] 'le basi materiali della significazione [1978]'" (Sebeok, "Foreword" in Capozzi, 1997, p. xiv).

exclusive concentration on the signs of intentional communication (semiology of communication). These reflected dominant trends in semiology following Saussure. Instead, semiotics after Sebeok is not only *anthroposemiotics* but also *zoösemiotics, phytosemiotics, mycosemiotics, microsemiotics, machine semiotics, environmental semiotics* and *endosemiotics* (the study of cybernetic systems inside the organic body on the ontogenetic and phylogenetic levels). All of this is under the umbrella of *biosemiotics* or, increasingly now and in the future, just plain *semiotics*.

In Sebeok's view, biological foundations, therefore, biosemiotics, form the epicenter of the study of both communication and signification in the human animal. In this perspective, the research of Jakob von Uexküll, biologist, teacher of Konrad Lorenz and one of the cryptosemioticians most studied by Sebeok, belongs to the history of semiotics.

Sebeok's semiotics unites what other fields of knowledge and human praxis generally keep apart either for justified exigencies of a specialized order, or because of a useless and even harmful tendency toward short-sighted sectoralization. Such an attitude is not free of ideological implications, which are often poorly masked by motivations of a scientific order.

Biology and the social sciences, ethology and linguistics, psychology and the health sciences, their internal specializations — from genetics to medical semiotics (symptomatology), psychoanalysis, gerontology and immunology — all find in semiotics, as conceived by Sebeok, the place of encounter and reciprocal exchange, as well as of systematization and unification. All the same, it must be stressed that systematization and unification are not understood here neopositivistically in the static terms of an "encyclopedia," whether this takes the form of the juxtaposition of knowledge and linguistic practices or of the reduction of knowledge to a single scientific field and its relative language (neopositivistic physicalism). Global semiotics may be presented as a *metascience* that takes all sign-related academic disciplines as its field. It cannot be reduced to the status of philosophy of science, although as a science it is engaged in dialogic relation with philosophy.

Sebeok reaches a global view through a continuous and creative shift in perspective that favors the development of new interdisciplinary relationships and new interpretive practices. Sign relations are identified where, for some, there seemed to be no more than mere "facts" and relations among things, independent from communication and interpretive processes. Moreover, this continual shift in perspective also favors the discovery of new cognitive fields and languages, which act dialogically. They are the dialogic interpreted-interpretant signs of fields and languages that already exist. As he explores the boundaries and margins of the sciences, Sebeok dubs this open nature of semiotics the "doctrine of signs."

2. The human capacity for modeling and the creation of new worlds

2.1. Language, modeling and the origin of signs
A Sign is a Just a Sign includes a paper of 1989, "Semiosis and Semiotics: What lies in Their Future?."[4] Here Sebeok significantly adds another meaning to the term "semiotics" understood as the general science of signs. This new meaning refers to the *specificity of human semiosis* and is of vital importance for a *transcendental founding of semiotics* as a doctrine of signs. Says Sebeok:

> Semiotics is an exclusively human style of inquiry, consisting of the contemplation — whether informally or in formalised fashion — of semiosis. This search will, it is safe to predict, continue at least as long as our genus survives, much as it has existed, for about three million years, in the successive expressions of Homo, variously labelled — reflecting, among other attributes, a growth in brain capacity with concomitant cognitive abilities — habilis, erectus, sapiens, neanderthalensis, and now s. sapiens. Semiotics, in other words, simply points to the universal propensity of the human mind for reverie focused specularly inward upon its own long-term cognitive strategy and daily manoeuvrings. Locke designated this quest as a search for 'humane understanding;' Peirce, as 'the play of musement.' (Sebeok, 1991b, p. 97)

This particular meaning of the term "semiotics" is obviously connected to "semiotics" understood as the general study of signs and of the typology of semiosis.

The exquisitely human propensity for musement implies the ability to carry out such operations as predicting the future or "traveling" through the past, that is, the ability to construct, deconstruct and reconstruct reality, inventing new worlds and interpretive models. The happy expression *The Play of Musement* is the title used by Sebeok, interpreter of Peirce, for his book of 1981.

In his paper "The Evolution of Semiosis," now included in *A Sign is a Just a Sign* (pp. 83-96), Sebeok explains existing correspondences between the various branches of semiotics and the different types of semiosis, from the world of micro-organisms to the Superkingdoms and the human world. Specific human semiosis, anthroposemiosis, is characterized as semiotics thanks to a modeling device specific to humans called by Sebeok "language." (We now know that *Homo habilis* was endowed with language, but not speech).

In Sebeok's research semiotics is interpreted and practiced as a life science, as biosemiotics. It follows that semiotics, as conceived by Sebeok, may be situated in a tradition of thought relating to the founders and masters of semiotics, including such figures as Hippocrates, Galen, Peirce, von Uexküll and in very recent times René Thom — an important Peirce scholar and topologist with competencies of a biological order.

In this perspective, Sebeok's semiotics examines the problem of the origin of signs. This is nothing less than the problem of the genesis of the universe (which

---

4. Originally written on invitation from Norma Tasca, representing the Associacao Portuguesa de Semiotica, for the Portuguese journal *Culture e Arte, 52*, 1989; now in Sebeok (1991b, pp. 97-99).

following Peirce is perfused with signs) from the free flow of energy-information to signals and signs.

The development of semiosis and its complex articulation coincides with the evolution of terrestial life from a single cell to its present-day multiform diversity, subdivided into three (or four) Superkingdoms: plants, animals and fungi. These kingdoms coexist and interact with the microcosm and together form the "biosphere." What Lotman calls the "semiosphere" to refer to the cultural dimension of signs, in reality coincides with the "biosphere" (see Sebeok, 1991b, p. 98), so that together the semiosphere and the biosphere form what we may call the great 'biosemiosphere.'

In another epochal paper included in *A Sign is a Just a Sign*, entitled "In what Sense is Language a 'Primary Modeling system'?" (also in Sebeok 1991a, pp. 175-196, and in a slightly modified version in Sebeok, 2001b, pp. 139-150), Sebeok describes language as a *modeling device*. Every species is endowed with a model that produces its own world, and language is the name of the model belonging to human beings. However, human language as a modeling device is completely different from the modeling devices of other life forms. The distinctive feature of human language is what the linguists call s*yntax*. Thanks to syntax hominids have not only one "reality," one world, but are also able to frame an indefinite number of possible worlds. This capacity is unique to human beings. Thanks to syntax human language is like Lego building blocks, it can reassemble a limited number of construction pieces in an infinite number of different ways. As a modeling device language can produce an indefinite number of models; in other words, the same pieces can be taken apart and put together to construct an infinite number of different models. Thanks to language, then, not only do human animals produce worlds as do other species, but, as Leibniz says, human beings can also produce an infinite number of possible worlds. This brings us back to the 'play of musement,' a human capacity which Sebeok considers particularly important for scientific research and all forms of investigation as well as for fiction and all forms of artistic creation.

Speech, like language, made its appearance as an adaptation, but f*or the sake of communication* and much later than language, precisely with *Homo sapiens*. Consequently, language too ended up becoming a communication device; and speech developed out of language as a derivative *exaptation* (this designation is proposed by Gould and Vrba, 1982). Exapted for communication, first in the form of speech and later of script, language enabled human beings to enhance the nonverbal capacity with which they were already endowed. On the other hand, speech was *exapted* for modeling and eventually functioned as a *secondary modeling system*. In addition to increasing the communication capacity, speech also increased the capacity for innovation and "play of musement." Such aspects as the plurality of languages and "linguistic creativity" (Chomsky, 1965, 1986) testify to the capacity of language understood as a primary modeling device, for producing numerous possible worlds.

The Forms of Meaning. Modeling Systems Theory and Semiotic Analysis by Sebeok and Marcel Danesi (2000) continues developing the fundamental notion of model as conceived in Sebeok's semiotics. Sebeok adapts the concept of modeling from the so-called Moscow-Tartu school (A. A. Zaliznjak, V. V. Ivanov, and V. N. Toporov. Ju. M. Lotman) where it is used to denote natural language (primary modeling system) as well as other human cultural systems (secondary modeling systems). However, differently to the Moscow-Tartu school, Sebeok goes further to extend the concept of modeling beyond the domain of anthroposemiosis. With reference to the biologist Jakob von Uexküll and his concept of *Umwelt* (J. von Uexküll, 1909) Sebeok's interpretation of model may be translated as "outside world model." On the basis of research in biosemiotics, the modeling capacity is observable in all life forms (see Sebeok, 1991b, pp. 49-58, 68-82; 1994a, pp. 117-127).

The study of modeling behavior in and across all life forms requires a methodological framework developed in the field of biosemiotics. This methodological framework is *modeling systems theory* proposed by Sebeok in his research on the interface between semiotics and biology. Modeling systems theory studies semiotic phenomena as modeling processes (Sebeok & Danesi, 2000, pp. 1-43). The applied study of modeling systems theory is called *systems analysis*, which distinguishes between primary, secondary and tertiary modeling systems.

In the light of semiotics conceived in terms of modeling systems theory, semiosis — a capacity with which all life forms are endowed — may be defined as "the capacity of a species to produce and comprehend the specific types of models it requires for processing and codifying perceptual input in its own way" (Sebeok & Danesi, 2000, p. 5).

2.2. The origin of language and speech
The question of the origin of human verbal language is often set aside by the scientific community as unworthy of discussion, having most often given rise to statements that are naïve and unfounded (an exception is offered in Fano, 1972, Eng. trans.: 1992). However, despite this general attitude Sebeok neither forgets the problem of the origins nor underestimates its importance. He claims that human verbal language is species-specific. It is on this basis that he interrogates — often with ironical overtones — the enthusiastic supporters of projects aimed at teaching captive primates verbal language. Sebeok points out the absurdity of such projects which are piloted by the false assumption that animals might be able to talk, or even more preposterous, that they possess the capacity for language understood as a syntactic modeling device. Sebeok's distinction between *language* and *speech* (See Sebeok, 1986, chpt. 2) not only protects against wrong-headed conclusions regarding animal communication, it also constitutes a general critique of phonocentrism and the tendency to base scientific investigation on anthropocentric principles.

According to Sebeok, language appeared and evolved as an *adaptation* much earlier than speech in the evolution of the human species to *Homo sapiens*. Language does not arise as a communicative device (a point on which Sebeok is in accord with Chomsky, even though the latter does not make the same distinction between *language* and *speech*); in other words, the specific function of language is not to transmit messages or to give information.

Instead, Sebeok describes language as a *modeling device*. Every species is endowed with a model that "produces" its own world, and language is the model belonging to human beings. However, as a modeling device, human language is completely different from the modeling devices of other life forms. Its characteristic trait is what the linguists call *syntax*, the ordering and operational rules of individual elements. But, while for linguists such elements ordered by syntax are words and phrases, instead Sebeok refers to a mute syntax when he speaks of syntax in language. Thanks to syntax, human language is like Lego building blocks. It can reassemble a limited number of construction pieces in an infinite number of different ways. As a modeling device, language can produce an indefinite number of models; in other words, the same pieces can be taken apart and put together to construct an infinite number of different models.

And thanks to language, not only do human animals produce worlds similarly to other species, but as also Leibniz said, human beings can produce an infinite number of possible worlds. This brings us back to the "play of musement," a human capacity that Sebeok following Peirce considers particularly important for scientific research and all forms of investigation, and not only for fiction and all forms of artistic creation.

Speech, like language, made its appearance as an adaptation, but *for the sake of communication* and much later than language, precisely with *Homo sapiens*. Speech organizes and externalizes language. Consequently, language too ended up becoming a communication device, enhancing the nonverbal capabilities already possessed by human beings; and speech developed out of language as what some evolutionary biologists call a derivative *exaptation* (see Gould & Vrba, 1982, pp. 4-15).

2.3. Iconicity and language
Sebeok believes that language as a modeling device relates iconically to the universe it models. This statement links him directly to Peirce and Jakobson, both of whom stressed the importance of iconic signs. An equally important connection can be made with Ludwig Wittgenstein's *Tractatus*, particularly with the notion of "picturing."

Wittgenstein begins his work on the processes that produce language-thought and on semiotic-cognitive procedures in his *Tractatus* (see Wittengenstein 1922). However, he subsequently sets aside this aspect of his research in *Philosophical Investigations* (see Wittgenstein 1953, 1967), where he focuses on meaning as use and on linguistic conventions (linguistic games). *Philosophical Investigations* is considered as an important turning point in Wittgenstein's research, especially by analytical philosophers, all the same we must not lose sight of his *Tractatus* and its importance, particularly as regards the iconic aspect of language (cf. Ponzio, 1991; 1997, pp. 309-313). In the *Tractatus*, Wittgenstein distinguishes between names and propositions: the relation between names or "simple signs" used in the proposition and their objects or meaning is of the conventional type. The relation between whole propositions or "propositional signs" and what they signify is a relation of similarity. The proposition is a logical picture (cf. *CP* 4.022, 4.026). As much as propositions are also conventional-symbolic, they are fundamentally based on the relation of representation, that is, the iconic relation; and, similarly to Peirce's "existential graphs," this relation is of the proportional or structural type.

The iconic relation can also be explained and analyzed through the distinction between *analogy, isomorphism* and *homology*, discussed by Ferruccio Rossi-Landi (1968/1992, 1983). The distinction between analogy and homology is congenial to the general orientation of Sebeok's own research, given its association with biology. His method is homological.

This approach to the relation between language and the world also has implications for the theory of knowledge, for the study of cognitive processes and psychology, which Sebeok directly addresses in terms of psycholinguistics and psychosemiotics. Relating semiotics to neuro-biology, he describes the mind as a sign system or model representing the surrounding world or *Umwelt*. The model is an icon, a kind of diagram, where the most pertinent relations are of a spatial and temporal order. These relations are not fixed once and for all but can be fixed, modified and fixed again in correspondence (a resemblance relation) with the *Innenwelt* (inner world) of the human organism. On the basis of this model which therefore is comparable to a diagram or a map, the human mind shifts from one node to another in the sign network, choosing each time the interpretive path considered most suitable (see Sebeok, 1986, Chpt. 7).

2.4. Living, lying and musing...

In Italy, long before Umberto Eco (1975, 1976; see Petrilli & Ponzio, in press, Part One, Chpt. VII) defined semiotics as the discipline that studies lying, Giovanni Vailati[5] had stated that signs may be used for deviating and deceiving. He entitled his review of Giuseppe Prezzolini's *L'arte di persuadere*, "Un manuale per bugiardi" (A handbook for liars). This particular aspect of Vailati's studies is analyzed by Augusto Ponzio in his 1988 monograph dedicated to the Italian philosopher and semiotician Ferruccio Rossi-Landi (see Petrilli & Ponzio, in press, Part One, VI; see also Ponzio 1990a, and 1990b). "Plurivocità, omologia, menzogna" (Plurivocality, homology, lying) is the title of a section included in a chapter dedicated to the relation between Rossi-Landi and Vailati, the former's predecessor. Sebeok himself also evokes Vailati in relation to Peirce in his paper "Peirce in Italia" of 1982. He describes the non-isomorphic character of signs with respect to reality and also analyzes the use of signs

---

5. Giovanni Vailati (1863-1909) was a mathematician, logician and pragmatist philosopher. He was a pupil of Giuseppe Peano, Vailati lectured in mathematics and physics at the University of Turin (in 1892 and 1899) and subsequently taught in various State schools in the North of Italy. He corresponded with such personalities as Franz Brentano and Victoria Welby whose Significs he appreciated and developed. He acknowledged the importance of Peirce's pragmatism that he introduced to the Italian intellectual scene. In his short lifetime he distinguished himself as an innovative thinker in philosophy of language, history of science, and epistemology (interpretation semiotics).

Independently from Peirce, Vailati was conscious of the importance of abduction for experimentation and discovery. In Italy the explicit and programmatic continuation of language studies in the direction indicated by Vailati is represented by the work of Ferruccio Rossi-Landi (see Vailati 1972, 1987, 2000; Petrilli 1990a; Ponzio, 1990b; Quaranta, 1989).

for lying (yet another leitmotif in his research) — that is, the use of signs for fraud, illusion and deception, the capacity of signs for masking and pretence.

Deception, lying, and illusion are forms of behavior which a semiotician like Sebeok, entranced by signs wherever they occur, could not resist. For example, he was attracted by the signs of the magician and constantly returned to forms of behavior and situations of the "Clever Hans" type. It was thought that the horse known as Clever Hans knew how to read and write, but in reality it was an able interpreter of the signals communicated by its trainer, either inadvertently or voluntarily in an intentional attempt at fraud (cf. Sebeok, 1979, pp. 85-106).

Sebeok explored the capacity for lying in the nonhuman animal world, which we believe to be an interest with a dual motivation. The first concerns Sebeok's commitment to contradicting the belief that animals can "talk" in a literal sense, that they are invested with a characteristic — language — which, on the contrary, is species-specific to humankind alone. In some cases, Sebeok's commitment to unveil this widespread mystification involves unmasking the fraudulent acts of impostors; in others it involves undermining illusions. With recourse to given theories, empirical documentation and even parody (cf. Sebeok, 1986, pp. 145-148), Sebeok has made an important contribution to evidencing the absurd, often ridiculous and no doubt scientifically unsound consequences of ignoring and abstracting from species-specific differences between human verbal language and animal language, or better sign systems.

The second motivation for Sebeok's inquiry into the capacity for lying in the nonhuman animal world is related to his wish to explore the fascinating question of whether nonhuman animals lie in the same way that humans do. As evidenced by studies in zoösemiotics, signs do not belong exclusively to the human world and it may well be that the use of signs also implies the ability to lie (cf. Sebeok, 1986, pp. 126-130).

3. Sebeok's works and the destiny of semiosis

3.1. A tetralogy
Over a decade (1976 to 1986) Sebeok, published the tetralogy formed by *Contributions to the Doctrine of Signs* (1976), *The Sign & Its Masters* (1979), *The Play of Musement* (1981), *I Think I Am a Verb* (1986).

In the opening lines to *The Sign & Its Masters* (see the programmatic chapters: 1, "Semiosis in Nature and Culture," pp. 3-26; and 4, "Ecumenicalism in Semiotics," pp. 61-84), Sebeok describes his book of 1979 as "transitional." In truth, this remark may be extended to the whole of his research if considered in the light of recent developments in philosophico-linguistic and semiotic debate. However, our allusion is to the transition from "code semiotics" (that is, the conception of the sign as a message that has been encoded and only calls for decodification, so that comprehension of the sign is reduce to mere decodification) to "interpretation semiotics," where code semiotics is centered on linguistics and, therefore, on verbal signs, while "interpretation semiotics," unlike the former, also accounts for the autonomy and arbitrariness of nonverbal signs, whether "cultural" or "natural."

In his survey of the problems relevant to semiotics and of the masters of signs, Sebeok discusses various aspects characterizing the "cultural" and "natural" approaches to semiotics, which may be most simply summarized with two names — Ferdinand de Saussure and Charles S. Peirce. The study of signs is "in transit" from "code semiotics" to "interpretation semiotics" as represented by these two emblematic figures, and in fact has now decidedly shifted in the direction of the latter.

Contributions to the Doctrine of Signs has a strong theoretical bias; here Sebeok had already expressed his preference for the semiotics of interpretation. *The Play of Musement*, a collection of papers published in 1981, explores the efficacy of semiotics as a methodological tool and the potential range of its application and does so in more discursive terms, although in both these books Sebeok's perspective has solid theoretical foundations. By contrast, *The Sign & Its Masters*, the in-between book, considers the different possibilities that branch out from our two semiotic alternatives, code semiotics and interpretation semiotics. In fact, in addition to being a compact theoretical book, *The Sign & Its Masters* also offers a survey of the various alternatives, positions and phases in sign studies as incarnated throughout history by important scholars of signs, whether they have dealt with signs either directly or indirectly.

Sebeok's position is distant from Saussure's who limited the sign science to the more restricted spaces of the signs of human culture, and still more reductively to signs produced intentionally for communication. Instead for Sebeok no aspect of sign life must be excluded from semiotics, no limits are acceptable on semiotics, whether contingent or deriving from epistemological conviction. At the same time, however, contrary to eventual first impressions, Sebeok's work discourages any claims to the status of scientific or philosophical omniscience; indeed there is no expectation to solve all problems indiscriminately.

Sebeok's writings transform us into the direct witnesses and interpretants of (abductive) turning points in his research as he experiments, discusses and evaluates different methods of semiotic inquiry, identifies possible objects of analysis and outlines the boundaries, or, better, suggests the boundlessness of semiotics as a disciplinary field. From this point of view *The Sign & Its Masters* — as, in reality, all of his research — is *transitional* insofar as it contributes significantly to the shift towards interpretation semiotics. This shift frees sign study once and for all from subordination to (Saussurean) linguistics and from false dichotomies: communication semiotics versus signification semiotics, referential semantics versus nonreferential semantics (see Eco, 1975).

3.2. Semiotics as a doctrine of signs and metasemiosis
Despite such a totalizing approach to semiotics Sebeok most significantly uses neither the ennobling term "science" nor the term "theory" to name it. Instead, as we have seen, he repeatedly favors the expression "doctrine of signs," adapted from John

Locke according to whom a doctrine is a body of principles and opinions that vaguely form a field of knowledge. Sebeok also uses this expression as understood by Charles S. Peirce (that is, with reference to the instances of Kantian critique). This is to say that Sebeok invests semiotics not only with the task of observing and describing phenomena, in this case signs, but also of interrogating the conditions of possibility that characterize and specify signs for what they are, as emerges from observation (necessarily limited and partial), and for what they must be (cf. Sebeok's "Preface" to *Contributions to the Doctrine of Signs*, 1976).

This humble and at the same time ambitious character of the "doctrine of signs" leads Sebeok to a Kantian critical interrogation of its very conditions of possibility: the doctrine of signs is the science of signs that questions itself, attempts to answer for itself, and inquires into its very own foundations. As a doctrine of signs, semiotics is also philosophy not because it deludes itself into believing that it can substitute philosophy, but because it *does not* delude itself into believing that the study of signs is possible without philosophical questions regarding its conditions of possibility.

3.3. From the nonhuman interpreter as a sign to the human interpreter as a verb
I Think I Am a Verb (1986) is the fourth book in Sebeok's tetralogy of the 1970s and 1980s. Since then other important volumes have followed in rapid succession. These include: *Essays in zoösemiotics* (1990b), *A Sign is Just a Sign* (1991b), *American Signatures* (1991a), *Semiotics in the United States* (1991c), *Signs. An Introduction to Semiotics* (2001b), *Come comunicano gli animali che non parlano* (1998b), *Global Semiotics* (2001b), without forgetting important earlier volumes such as *Perspectives in zoösemiotics* (1972), plus numerous others under his editorship including *Animal Communication* (1968), *Sight, Sound, and Sense* (1978), and *How Animals Communicate* (1979).

Rather than continue this long list of publications, it will suffice to remember that Sebeok has been publishing since 1942. His writings are the expression of ongoing research and probing reflection over more than half a century as he interprets the semiosic universe, whose infinite multiplicity, variety and articulation he has substantially contributed to manifesting.

I Think I Am a Verb is a book which at once assembles a broad range of interests and also acts as a launching pad for new research itineraries in the vast region of semiotics. The title evokes the dying words of the eighteenth President of the United States, Ulysses Grant, which ring with Peircean overtones. In fact, in Peirce's view, man is a sign just as all living beings are. However, Sebeok's choice of a verb instead of a noun to characterize this sign serves to emphasize the condition of continuous becoming, transformation and renewal of signs in the human world.

A fundamental point in Sebeok's doctrine of signs is that living is sign activity. To maintain and to reproduce life, and not only to interpret it at a scientific level, are all activities that necessarily involve the use of signs. Sebeok theorizes a direct connection between the biological and the semiosic universes, and, therefore, between biology and semiotics. His research would seem to develop Peirce's conviction that man is a sign with the addition that this sign is a verb: to interpret. And in Sebeok's particular conception of reality, the interpreting activity coincides with the life activity, and in his own personal case, with the whole of his life. If I am a sign, as he would seem to be saying through his life as a researcher, then nothing that is a sign is alien to me — *nihil signi mihi alienum puto*; and if the sign situated in the interminable chain of signs is necessarily an "interpretant"— the term Peirce gave to the effect of a sign, an effect that is itself a sign — then "to interpret" is indeed the verb that may best help me understand who I am.

3.4. European and American semiotics, a dialogue
In *Semiotics in the United States*, Sebeok analyzes U.S. semiotics at three different levels, at once closely interrelated and yet easily identifiable.

At the *first* level he makes both a synchronic and diachronic survey of the various theoretical trends, perspectives, problems, fields, specializations and institutions that characterize U.S. semiotics. Regarding the diachronic perspective, Sebeok takes on the difficult task of reconstructing the origins of American semiotics. To this end he researches discourse that was not yet connoted as semiotics at the time and that in certain cases is still today considered marginal with respect to semiotics or completely distant from it.

The *second* level is theoretical and critical. Sebeok takes a stand with respect to given problems in semiotics: problems of a general order concerning, for instance, the delimitation of the field of semiotics or the construction of a general sign model; and problems of a more specific order concerning the various sectors and subsectors of the science, or "doctrine of signs." The impression that Sebeok would seem to confirm here and there is that this more problematic level sets the perspective for the whole volume: it completes the first level and avoids limiting the volume to pure historical descriptivism.

The *third* level is connected to the second in the sense that while developing and illustrating his theoretical views, Sebeok colors them with personal overtones and most often with amusing biographical anecdotes. There are very few pages in *Semiotics in the United States* where Sebeok does not figure as one of the characters populating the stories, episodes and enterprises forming his narration. In fact, this is largely due to his surprising and perhaps unprecedented involvement in the organization and promotion of the semiotic science at a world level — a cause to which he has been committed since the gradual emergence of semiotics as a discipline in its own right. Sebeok has been in direct contact with many of the authors mentioned in his volume and has many "memories" of personal experiences with them. Consequently these memories have found their way into his description of the problems and orientations characterizing the semiotic globe.

With reference to these three shaping factors another book by Sebeok which recalls *Semiotics in the United States* is the oft-cited *The Sign & Its Masters*. Here in fact the historical, theoretical-critical and anecdotal threads of Sebeok's discourse converge and interweave even more than in his other books, though the autobiographical aspect is never lacking in any one of them. *Semiotics in the United States* may also be related to *I Think I Am a Verb* where autobiographical motivations are not lacking in the choice of topics, authors and personalities cited, including the eighteenth President of the United States of America, Ulysses S. Grant, whose dying words as we have said inspire the title of the volume.

Sebeok's interests cover a broad range of territories ranging from the natural sciences to the human sciences. He deals with theoretical issues and their applications from as many angles as are the disciplines called in question: linguistics, cultural anthropology, psychology, artificial intelligence, zoology, ethology, biology, medicine, robotics, mathematics, philosophy, literature, narratology and so forth. Even though the initial impression might be of a rather erratic mode of proceeding as he experiments various perspectives and embarks upon different research ventures, in reality Sebeok's expansive and seemingly distant interests find a focus in his "doctrine of signs" and in the fundamental conviction subtending his general method of enquiry, that is, the universe is perfused with signs, indeed, as Peirce hazards, may be composed exclusively of signs.

Through his numerous publications Sebeok has promoted a wide-ranging vision of semiotics which coincides with the study of the evolution of life. After Sebeok's work both our conception of the semiotic field and of the history of semiotics have changed noticeably. Thanks to him semiotics at the beginning of the new millennium is proposing a radically broader view than that presented during the first half of the 1960s.

What is immediately striking about Sebeok's work may be described as his "dialogic" and "polyphonic" approach (in the Bakhtinian sense of these words). Sebeok promotes dialogue among signs, among the different orders of signs, among different interpretive practices, domains and fields, as well as among the "masters" of signs, including those who had never previously been in direct contact with each other, or who did not even suspect they were dealing with signs (his so-called "cryptosemioticians").

As is evident throughout his thought system, Peirce too recognized the importance of "dialogism" in the development of thought, and even more broadly in the evolution of life generally, of which human thought processes are a part. In a letter to Victoria Welby of 2nd December 1904, and very much in accord with her own views, Peirce, who had been forced into isolation after having been excluded from academic life, wrote that "after all a philosophy can only be passed from mouth to mouth, where there is opportunity to object & cross-question" (Hardwick, 1977, p. 44).

As testified by his long teaching career and constant commitment to promoting the "community of inquirers," Sebeok believed continuity in dialogic exchange is nothing less than of vital importance. Indeed, as Iris Smith states in her introduction to Sebeok's book of 1991, *American Signatures: Semiotic Inquiry and Method*, his own peculiar way of living his condition as an intellectual testifies to the fact that individual reflection must be measured against the reflection of others.

We believe that Sebeok's awareness of the vastness, variety and complexity of the territories he explores and of the problems analyzed demonstrates a sense of utmost prudence, sensitivity to problems and humility in the interpretations he offers. This is the case not just when he ventures over the treacherous territory of signs, but still more in relation to the deceptive sphere of the signs of signs — the place of his semiotic probings.

3.5. The destiny of semiosis after life

Semiosis extends over all terrestrial biological systems, from the sphere of molecular mechanisms at the lower limit, to a hypothetical entity at the upper limit christened "Gaia," the Greek for "Mother Earth" — a term introduced by scientists toward the end of the 1970s to designate the whole terrestrial ecosystem that englobes the interactive activity of different forms of life on Earth.

As Sebeok says, alluding to the fantastic worlds of *Gulliver's Travels*, semiosis spreads over the Lilliputian world of molecular genetics and virology to Gulliver's man-size world, and finally to the world of Brobdingnag, Gaia, our gigantic bio-geo-chemical ecosystem.

And beyond? Can we assert that semiosis extends beyond Gaia? A "beyond" understood in terms of space, but also of time? Is semiosis possible beyond Gaia, outside it, and beyond this gigantic organism's life span? Sebeok ponders this question to (see "Semiosis and Semiotics. What lies in their future?" In Sebeok, 1991b, p. 98).

With his research Sebeok takes stock of the impressive general progress and expansion of the semiotic field during approximately the past twenty to thirty years. Starting from an oversimplifying definition of semiotics as the study of the exchange of any kind of message and related sign systems (which he criticized), he theorizes semiotics as the "play of musement" mediating between reality and illusion:

> the central preoccupation of semiotics is an illimitable array of concordant illusions; its main mission to mediate between reality and illusion — to reveal the substratal illusion underlying reality and to search for the reality that may, after all, lurk behind that illusion. This abductive assignment becomes, henceforth, the privilege of future generations to pursue, insofar as young people can be induced to heed the advice of their elected medicine men. (Sebeok, 1986, pp. 77-78)

We believe that the question posed by Sebeok concerning the destiny of semiosis also derives from awareness of the responsibility of semiotics relatively to semiosis. Going beyond Sebeok we now believe that the time has come for global semiotics to evolve into what we propose to call "semioethics" (see Petrilli & Ponzio, in press, Part Three; Ponzio & Petrilli, 2003).

References

Chomsky, N. (1965). *Aspects of a theory of syntax*. Cambridge, MA: The MIT Press.
Chomsky, N. (1986). *Knowledge of language: Its nature, origin, and use*. New York: Praeger.
Eco, U. (1975). *Trattato di semiotica generale*. Milan: Bompiani.

Eco, U. (1976). *A theory of semiotics*. Bloomington, IN: Indiana University Press. (Original work published in Italian in 1975.)
Fano, G. (1972). *Origini e natura del linguaggio*. Turin: Einaudi.
Fano, G. (1992). *The origins and nature of language*. (Eng. trans. Ed. and Intro. by S. Petrilli.) Bloomington: Indiana University Press, 1992. (Original work published in 1972.)
Gould, S. J. & Vrba, E. (1982). Exaptation. *Paleobiology 8* (1), 14-15.
Hardwick, C. S. (Ed. with the assistance of J. Cook) (1977). *Semiotic and significs. The correspondence between Charles S. Peirce and Victoria Lady Welby*. Bloomington: Indiana University Press.
Morris, C. (1946). *Signs, language and behavior*. New York: Prentice Hall.
Morris, C. (1949). *Segni, linguaggio e comportamento* (S. Ceccato, Trans.). Milan: Longanesi. (Original work published in 1946.)
Morris, C. (1971). *Writings on the general theory of signs* (T. A. Sebeok, Ed.). The Hague: Mouton.
Ogden, C. K., & Richards, I. A. (1923). *The meaning of meaning*. London: Routledge and Kegan and Paul. (Also New York: Harcourt Brace Jovanovich, 1989)
Peirce, C. S. (1931-1966). *Collected papers of Charles Sanders Peirce* (Vols. 1-8, C. Hartshorne, P. Weiss, & A. W. Burks, Eds.). Cambridge, MA: Harvard University Press. (Citations in text follow this convention: *CP*, volume: paragraph number(s).)
Petrilli, S. (1990). The problem of signifying in Welby, Peirce, Vailati, Bakhtin. In A. Ponzio, *Man as a sign: Essays on the philosophy of language* (pp. 315-363). Berlin: Mouton de Gruyter.
Petrilli, S., & Ponzio, A. (In press). *Semiotics unbounded: Interpretive routes through the open network of signs*. Toronto: Toronto University Press.
Petrilli, S. & Ponzio, A. (2001). *Thomas Sebeok and the signs of life*. Cambridge: Icon Books.
Petrilli, S., & Ponzio, A. (2002). Sign vehicles for semiotic travels: Two new handbooks. *Semiotica 141*(1/4), 203-350.
Ponzio, A. (1988). *Rossi-Landi e la filosofia del linguaggio*. Bari: Adriatica.
Ponzio, A. (1990a). *Man as a sign: Essays on the philosophy of language* (S. Petrilli, Trans. & Ed). Berlin: Mouton de Gruyter.
Ponzio, A. (1990b). Theory of meaning and theory of knowledge: Vailati and Lady Welby. In H. W. Schmitz (Ed.), *Essays in Significs* (pp. 165-178). Amsterdam: John Benjamins.
Ponzio, A. (1991). Segno e raffigurazione in Wittgenstein. In A. Ponzio, *Filosofia del linguaggio 2: Segni valori ideologie* (pp. 185-202). Bari: Adriatica.
Ponzio, A. (1997). *Che cos'è la letteratura?* Lecce: Milella.
Ponzio, A., & Petrilli, S. (2001). *Il sentire della comunicazione globale*. Rome: Meltemi.
Ponzio, A., & Petrilli, S. (2003). *Semioetica*. Rome: Meltemi.
Posner, R.; Robering, K., &. Sebeok, T. A. [Eds.]. (1997-98). *Semiotik semiotics. A handbook on the sign-theoretic foundations of nature and culture* (Vols. 1-3). Berlin: Walter de Gruyter. (Vol. 3 is forthcoming).
Prezzolini, G. (1907). *L'arte di persuadere*. Florence: Lumachi.
Prodi, G. (1977). *Le basi materiali della significazione*. Milan: Bompiani.
Quaranta, M. (Ed.) (1989). *Giovanni Vailati e la cultura del '900*. Bologna: Arnaldo Forni.
Rossi-Landi, F. (1992). *Il linguaggio come lavoro e come mercato*. Milan: Bompiani. (Reprint of 1968 work).
Rossi-Landi, F. (1983). *Language as work and trade* (M. Adams et al., Trans.). South Hadley, MA: Bergin and Garvey (Original work published in 1968.)
Saussure, F. de (1916). *Cours de linguistique générale* (C. Bally and A. Séchehaye, Eds.). Paris: Payot.
Sebeok, T. A. (Ed.). (1968). *Animal communication: Techniques of studies and results of research*. Bloomington, IN: Indiana University Press.
Sebeok, T. A. (1972). *Perspectives in zoösemiotics*. The Hague: Mouton.
Sebeok, T. A. (1976). *Contributions to the doctrine of signs*. Bloomington, IN: Indiana University Press.
Sebeok, T. A. (Ed.). (1978). *Sight, sound and sense*. Blooomington, IN: Indiana University Press.
Sebeok, T. A. (1979a). *Contributi alla dottrina dei segni* (M. Pesaresi, Trans.). Milano: Feltrinelli. (Original work published in 1976)
Sebeok, T. A. (1979b). *The sign & its masters*. Austin, TX: The University of Texas Press.
Sebeok, T. A. (1981). *The play of musement*. Bloomington, IN: Indiana University Press.
Sebeok, T. A. (1982). Peirce in Italia. *Alfabeta, 35*, 1.
Sebeok, T. A. (1985). *Il segno e i suoi maestri* (S. Petrilli, Trans., Ed., and Intro). Bari: Adriatica.
Sebeok, T. A. (1986). *I think I am a verb*. New York: Plenum Press.
Sebeok, T. A. (1990a) *Penso di essere un verbo* (S. Petrilli, Trans., Ed. and Intro.) Penso di essere un verbo. Palermo: Sellerio, 1990. (Original work publshed in 1986).
Sebeok, T. A. (1990b). *Essays in zoösemiotics* (M. Danesi, Ed.). Toronto: University of Toronto Press.
Sebeok, T. A. (1991a). *American Signatures: Semiotic Inquiry and Method* (I. Smith,Ed.) Norman, OK: University of Oklahoma Press.
Sebeok, T. A. (1991b). *A sign is just a sign*. Bloomington, IN: Indiana University Press.
Sebeok, T. A. (1991c). *Semiotics in the United States*. Bloomington, IN: Indiana University Press.
Sebeok, T. A. (1992). *Sguardo sulla semiotica americana* (S. Petrilli, Trans. & Ed.). Milan: Bompiani. (Original work published in 1991.)

Sebeok, T. A. (1994a). Global Semiotics. In I. Rauch & G. F. Carr (Eds.), *Proceedings of the Vth International Congress of the International Association for Semiotic Studies: Semiotics around the world: Synthesis in diversity* (pp. 105-130). Berlin: Mouton de Gruyter. (Also in Sebeok 2001).
Sebeok, T. A. (1997). Foreword. In R. Capozzi, *Reading Eco. An anthology* (pp. xi-xvi). Bloomington, IN: Indiana University Press.
Sebeok, T. A. (1997-1998). The Evolution of Semiosis. In R. Posner, K. Robering,&. T. A. Sebeok, (Eds.). *Semiotik semiotics. A handbook on the sign-theoretic foundations of nature and culture* (Vols. 1, pp. 83-96). Berlin: Walter de Gruyter.
Sebeok, T. A. (1998a). *A sign is just a sign: La semiotica globale* (S. Petrilli, Trans. Ed. and Intro.). Milan: Spirali. (Original work published in English in 1991.)
Sebeok, T. A. (1998b). *Come comunicano gli animali che non parlano* (S. Petrilli, Trans., Ed., Intro.). Bari: Edizioni dal Sud.
Sebeok, T. A. (2001). *Global semiotics*. Bloomington: Indiana University Press.
Sebeok, T. A. (2001b). *Signs: An introduction to semiotics* (Rev. ed.). Toronto: Toronto University Press.
Sebeok, T. A., & Danesi, M. (2000). *The forms of meaning: Modeling systems theory and semiotic analysis*. Berlin: Mouton de Gruyter.
Sebeok, T. A., Petrilli, S., & Ponzio, A. (2001). *Semiotica dell'io*. Rome: Meltemi.
Sebeok, T. A., & Umiker-Sebeok, J. (Eds.). (1992). *Biosemiotics: The semiotic web 1991*. Berlin: Mouton De Gruyter.
Uexküll, J. von (1909). *Umwelt und Innenwelt der Tiere*. Berlin: Verlag von Julius Springer.
Uexküll, T. von (1997-1998). Biosemiose. In R. Posner, T. A. Sebeok, & K. Robering (Eds.), *Semiotics. A handbook on sign-theoretic foundations of nature and culture* (Vol. 1, pp. 447-456). Berlin: Walter de Gruyter.
Vailati, G. (1971). *Epistolario 1891-1909* (G. Lanaro, Ed.). Turin: Einaudi.
Vailati, G. (1972). *Scritti filosofici* (G. Lanaro, Ed.). Naples: Fulvio Rossi.
Vailati, G. (1987). *Scritti* (Vols. 1-3, M. Quaranta, Ed.) Sala Bolognese: Arnaldo Forni.
Vailati, G. (2000). *Il metodo della filosofia: Saggi di critica del linguaggio* (F. Rossi-Landi & A. Ponzio Eds.). Bari: Graphis.
Wittgenstein, L. (1922).*Tractatus logico-philosophicus* (Trans, by D. F. Pears and B. F. Guinness. Intro. by B. Russell.) London: Routledge & Kegan Paul.
Wittgenstein, L. (1953). *Philosophical investigation* (G.E.M. Anscombe, Trans.) Oxford: Blackwell.
Wittgenstein, L. (1967). *Ricerche filosofiche* (M. Piovesan & M. Trinchero, Eds.). Turin: Einaudi. (Original work published in 1953.)

Thomas A. Sebeok's Global Semiotics:
Modeling, communication, and dialogism

Augusto Ponzio[1]

In this paper we deal with Thomas A. Sebeok's conception of modeling and interrelation of semiosis in its entirety over the planet Earth. We develop his conception of interrelation through Mikhail M. Bakhtin's conception of dialogism. What connects Sebeok's global interrelation and Bakhtinian dialogism is their common interest in biology. Not only is Sebeok's semiotics based on the study of biology, but also Bakhtin's conception of dialogue, as demonstrated by the latter's commitment to writing a paper on contemporary vitalism in biology, published under the name of the biologist Ivan I. Kanaevon Through Bakhtin, and using Sebeok's global approach to semiotics, it is possible to extend dialogism beyond the sphere of anthroposemiosis to found all communication processes not only on modeling, as says Sebeok, but also on dialogism. This does not contradict the Peircean matrix of Sebeok's semiotics, since the concept of dialogue plays a fundamental role in the theory of Charles S. Peirce as well.

1. Modeling systems theory and global semiotics

Two pivotal topics in Thomas A. Sebeok's "doctrine of signs" — as he prefers to qualify semiotics instead of "science" or "theory of signs" — are modeling and interrelation between species-specific semioses over the entire planet Earth. The present paper focuses on these two topics developing them in the light of our own personal perspective. What we call *dialogism* in the present context refers precisely to the development of the situation of interrelation in global semiosis according to Sebeok. In our opinion modeling and dialogism are the basis of all communication processes. Another pivotal topic in Sebeok's semiotics is his belief that only humans are capable of "semiotics," that is, of consciousness, semiosic activity. We believe that this means that only human animals are responsible for semiosis in its entirety over the planet. And, given that in Sebeok's view semiosis and life coincide, the human animal is held responsible for the whole of life over the planet. Together with Susan Petrilli we have treated this question in other writings and specifically in our most recent book *Semioetica*, (Ponzio & Petrilli, 2003). Here, instead, we will concentrate on the problem of the connection among modeling, communication and dialogism.

After Sebeok semiotics is emerging as "global semiotics." According to the global semiotic perspective signs and life coincide and semiosis is behavior among living beings. A pivotal notion in global semiotics is that of "modeling" used to explain life

---

1. Augusto Ponzio is Full Professor of Philosophy of Language and Head of the Dipartimento di Pratiche linguistiche e analisi di testi, University of Bari. Mailing address: Dipartimento di Pratiche linguistiche e analisi di testi, Facoltà di Lingue e Letterature Straniere, Università di Bari, Via Garruba n. 6, Bari-70100, Italy. Home page: www.lingue.uniba.it/dip-plat/indice; Email: a.ponzio@lingue.uniba.it.

and behavior among living entities conceived in terms of semiosis. Therefore global semiotics also involves modeling systems theory.

Modeling is the foundation of communication. Communication necessarily comes about within the limits and according to the characteristics of the world which a given species models. Jakob von Uexküll speaks of invisible worlds to indicate the domain in which all animals are englobed according to which species they belong. What an animal perceives, craves, fears and hunts is relative to its own world. Human communication is the most complex and varied form of communication since man is that animal which is capable of modeling multiple possible worlds. Sebeok adapts the concept of modeling from the so-called Moscow-Tartu school though he enriches it by connecting it to the concept of *Umwelt* as formulated by Jakob von Uexküll (see Sebeok, 1991, pp. 49-58, 68-82; 1994, pp. 117-127; Sebeok & Danesi, 2000, pp. 1-43).

2. Semiotics and Semiosis

Sebeok's global semiotics is more than just a study of semiosis, it also carries out precise functions regarding semiosis. The globality of semiotics does not only concern its capacity for an overall view, but also its disposition for an overall response.

2.1. Three aspects of the unifying function of semiotics
As emerges from Sebeok's research, the unifying function of semiotics may be considered from the viewpoint of three strictly interrelated aspects all belonging to the same interpretive practice, highly characterized abductive creativity:
1) The descriptive-explanatory aspect
Semiotics singles out, describes and explains signs, that is, interpreted-interpretant relationships, forming events
   a) which are connected by a relation of contiguity and causality (indexical relation) and which therefore are given immediately and necessarily;
   b) or which, on the contrary, despite the distance between them on an indexical level are associated on the basis of a hypothesized, iconic relation of similarity;
     (b1) in some cases the iconic relation mainly results from obeying certain conventions (the iconic-symbolic relation);
     (b2) in other cases the iconic relation mainly results from the tendency to innovation (the iconic-abductive relation), and not from obeying prefixed convention. These interpreted-interpretant relations are identified not only in thematized objects, but also in the interpretive practices of different sciences.
Therefore the descriptive-explanatory function of semiotics is also practised in relation to cognitive processes themselves, in terms of critique in a Kantian sense, of the search for a priori possibilities or conditions.

2) The methodological aspect

Semiotics is also the search for methods of inquiry and acquisition of knowledge, both ordinary and scientific knowledge. From this point of view, and different from the first aspect, semiotics does not limit itself simply to describing and explaining, but it also makes proposals concerning cognitive behavior. Hence, semiotics overcomes the tendency to parochial specialism among the sciences when this causes separation from each other.

3) The ethical aspect

For this aspect we propose such terms as "ethosemiotics" or "teleosemiotics" (from "telos" = end) or, better, "semioethics." In this aspect, the unifying function of semiotics concerns proposals and practical orientations for human life in its wholeness (human life considered in all its biological and socio-cultural aspects). The focus is on what may be called the "problem of happiness." This problem is evidently considered to be very important by Herodotus, who early in the first book of the *Histories* narrates the downfall of the last of the Lydian kings, Croesus, who imagined himself to be the happiest of men.

In turn, the story of Croesus as described by Herodotus is interpreted by Sebeok. Happiness is impossible for Croesus to maintain because of his inability to appreciate each of his two sons: one endowed with the word, the other being deaf and dumb, and unnamed.

Sebeok's study "The Two Sons of Croesus: A Myth about Communication in Herodotus" (in Sebeok, 1979), reflects on this third aspect of semiotics which refers to the problem of wisdom as entrusted to myths, popular tradition and literature in particular genres (those described by Mikhail Bakhtin as belonging to "carnivalized literature," which derive from popular culture). By analogy with the deaf and dumb son of Croesus, we may remember King Lear's reticent Cordelia, or in *The Merchant of Venice*, the "muteness" and simplicity of the leaden casket — which contrary to common expectation is a sign that holds Portia's image.

Concerning this third aspect of the unifying function of semiotics, particular attention is paid to recovering the connection with what is considered and experienced as being separate. In today's world, the logic of production and the rules that govern the market allowing all to be exchanged and commodified threaten to render humanity ever more insensible to nonfunctional and ambivalent signs. These signs may range from the vital signs forming the body to the seemingly futile signs of phatic communication with others. Reconsideration of these signs and their relative interrelations would seem absolutely necessary in the present age to improve the quality of life. The economics of capitalist globalization imposes ecological conditions in which communication between ourselves and our bodies, as well as the environment, is ever more difficult and distorted (cf. Sebeok's considerations in Appendix I, "The Semiotic Self," in *The Sign and Its Masters* (1979); see also Sebeok, Petrilli & Ponzio, 2001). This third aspect of semiotics also operates in a way that unites rational worldviews with myth, legend, fable and all other forms of popular tradition concerned with the relationship of humans to the world about them. Such a function is rich with implications for human behavior: those signs of life that today we cannot or do not wish to read, and those signs that we do not know how to read may well recover one day their importance and relevance for humanity.

Quite frequently the study of sign function has been thought to be sufficient for an understanding of the nature of signs. On the contrary, Sebeok draws attention to the functioning of signs as an end in itself, which represents a sort of excess with respect to the function and purpose of signs. This excess is visible, for example, in ritual behavior among human beings and animals, but also in language. In fact beyond its communicative function, language may be considered as a sort of game without which such things as imagination, fantasy, or highly abductive reasoning would never be possible (for these aspects, cf. Sebeok's *The Play of Musement*, 1981).

2.2. Semiosis and semiotics. 'Semiotics', another meaning
In addition to the general science of signs Sebeok most significantly adds another meaning to the term *semiotics*, that is, as indicating, *the specificity of human semiosis*. This concept is clearly proposed in a paper of 1989, "Semiosis and semiotics: what lies in their future?" now chapter 9 in *A Sign Is Just a Sign* (Sebeok, 1991, pp. 97-99), and is of vital importance for a *transcendental founding of semiotics* given that it explains how semiotics as a science and metascience is possible. He writes:

> Semiotics is an exclusively human style of inquiry, consisting of the contemplation — whether informally or in formalized fashion — of semiosis. This search will, it is safe to predict, continue at least as long as our genus survives, much as it has existed, for about three million years, in the successive expressions of Homo, variously labeled — reflecting, among other attributes, a growth in brain capacity with concomitant cognitive abilities — *habilis, erectus, sapiens, neanderthalensis*, and now *s. sapiens*. Semiotics, in other words, simply points to the universal propensity of the human mind for reverie focused specularly inward upon its own long-term cognitive strategy and daily maneuverings. Locke designated this quest as a search for "humane understanding;" Peirce, as "the play of musement." (Sebeok, 1991, p. 97)

In his article "The evolution of semiosis" (in Posner, Robering, and Sebeok, Vol. I, 1997-98), Sebeok explains the correspondences that exist between the branches of semiotics and the different types of semiosis, from the world of micro-organisms to big kingdoms and the human world. Specific human semiosis, anthroposemiosis, is represented as semiotics thanks to the specifically human "modeling device" called *language*. This observation is based on the fact that it is virtually certain that *Homo habilis* was endowed with language, but not speech. Sebeok's distinction between language and speech corresponds, even if roughly, to the distinction between *Kognition* and *Sprake* drawn by Muller (1987) in *Evolution, Kognition and Sprake* (see Sebeok in Posner, Robering, & Sebeok, 1997-1998, Vol. I, p. 443).

3. Dialogism, Modeling and Communication in Semiosis

3.1. Model and modeling

As stated above, a fundamental concept in Sebeok's global semiotics is that of *model* which he develops from the so-called Moscow-Tartu school (see Lucid, 1977; Rudy, 1986). Here the concept of modeling is limited to the human sphere (Lotman's semiosphere) and distinguishes between the expression "primary modeling system," used to denote natural language, and "secondary modeling system," used for all other human cultural systems. Instead, Sebeok extends the concept of model beyond the domain of anthroposemiotics, connecting it to the research of the biologist Jakob von Uexküll and his concept of *Umwelt* which, in Sebeok's interpretation, may be translated as "outside world model." On the basis of research in biosemiotics, we now know that the modeling capacity is operative in all life forms. Semiosis may be interpreted as the capacity with which all life forms are endowed to produce and comprehend the species-specific models of their worlds (see Sebeok & Danesi, 2000, p. 5). Primary modeling is the innate capacity of organisms for simulative modeling in species-specific ways. The primary modeling system of the species *Homo* is *language* which should not be confused with verbal language, as does the Moscow-Tartu school. Verbal language is a system of communication, instead language is a species-specific modeling device. Consequently secondary and tertiary modeling systems presuppose language, so that they too are uniquely human capacities. Therefore, in Sebeok's terminology the secondary modeling system is verbal language or, *speech*, while tertiary modeling systems are all human cultural systems, symbol-based modeling processes grounded in language and speech. Sebeok's tripartite distinction is very important to distinguish between modeling and communication and to demonstrate the foundational character of modeling with respect to communication.

The tripartition of semiosis as proposed by Thure von Uexküll also contributes in this sense. However, as we shall now see in what follows, Uexküll's tripartition calls for "translation" from his own terminology belonging to code semiotics (a mixture of Saussurean semiology and Shannon and Weaver's information theory) with his introduction of such terms as "emitter" and "receiver" into the language of Peircean interpretation semiotics. This translation is pivotal in our understanding of the connection between modeling and dialogism.

To this end, it is also necessary to deal with the *Semiosic Matrix* as proposed by Martin Krampen.

Finally, another indispensable passage in our argumentation — developing Sebeok and beyond him — on the relation between modeling and dialogism as the foundation of communication is the *Functional Cycle* as described by Jakob von Uexküll.

All these aspects will now be dealt with in the following sections and eventually lead into our treatment of dialogism as conceived by Bakhtin and its possible relation to Sebeok's biosemiotics.

3.2. Reformulating Thure von Uexküll's typology of semiosis

In the article "Biosemiosis" (in Posner, Robering, and Sebeok 1997-98, Vol. I, Chapter III, pp. 447-456; see also "Varieties of Semiosis" by Uexküll in Sebeok and Umiker-Sebeok, 1992, pp. 455-470), Thure von Uexküll distinguishes between three different kinds of semiosis which are characterized by differences in the role carried out by the emitter and receiver. He calls these three kinds of semiosis: 1) *semiosis of information or signification*; 2) *semiosis of symptomatization*; and 3) *semiosis of communication*.

In *semiosis of information, or signification*, we have an inanimate environment which acts as a "quasi-emitter" without a semiotic function. The receiver, that is to say, a living entity, a living system, which makes whatever it receives meaningful via its receptors, must perform all semiotic functions.

In *semiosis of symptomatization* the emitter is a living being sending out signals through its behavior or attitude, which are not directed toward a receiver and do not wait for an answer. Signals as received by the receiver are signs called *symptoms*.

In *semiosis of communication* signs are emitted for the receiver and must find the meaning intended by the emitter (See Thure von Uexküll in Posner, Robering, and Sebeok 1997-98, Vol. I, Chapter III, pp. 449-450).

In our terminology, and in accordance with Peirce (but also with the terminology used by Thure von Uexküll in his approach to biosemiotics), these three kinds of semiosis which are specified in terms of the different roles carried out by the emitter and the receiver, may instead be reformulated in terms of the different roles carried out by the interpretant sign and the interpreted sign. We may state that:

a) the *interpreted* becomes a sign only because it receives an interpretation from the interpretant which is a response (*semiosis of information*); or

b) before it is interpreted as a sign by the interpretant, the interpreted is already itself an interpretant response (*symptom*) which, however, is not oriented to being interpreted as a sign (*semiosis of symptomatisation*);

c) before being interpreted as a sign by the interpretant, the interpreted is already an interpretant response oriented for interpretation as a sign, in other words, the interpreted calls for another interpretant response (*semiosis of communication*).

We believe that our reformulation of Thure von Uexküll's typology of semiosis, which distinguishes between the way the interpreted sign and the interpretant sign participate in interpretation, presents the following advantages over the conception of semiosic differences established on the basis of 'emitter' and 'receiver' participation:

a) the role of the interpretant in semiosis is emphasized;

b) the meaning of the expression *inanimate quasi-interpreter* in *semiosis of information or signification* is explained as the "interpreted-non-interpretant" (while in *semiosis of symptomatisation* the interpreted is an interpretant-interpreted which is not oriented to being interpreted as a sign; and in *semiosis of communication* the interpreted is an interpretant-interpreted which is oriented to being interpreted as a sign);

c) semiosis is identified with the capacity for interpretation, that is to say, for response;

d) importance of the pragmatic dimension in semiosis is confirmed;
e) Thure von Uexküll's definition of biosemiotics as *interpretation of interpretation* or, in a word, *metainterpretation* is also confirmed and developed.

In our reformulation we employ the same terminology used by Thure von Uexküll to describe his model of biosemiotics (Thure von Uexküll in Posner, Robering, and Sebeok 1997-98, Vol. I, Chapter III, p. 456).

Semiosis of information or signification, semiosis of symptomatization and semiosis of communication are grounded in specific types of modeling characteristic of specific life forms. The species's capacity for modeling is the necessary a priori for processing and interpreting perceptual input in its own species-specific way.

3.3. From 'substitution' to 'interpretation'

According to Sebeok (1994, pp. 10-14), both the Object (O) and the Interpretant (I) are Signs. Consequently, we may rewrite O as Son and I as SIn, so that both the first distinction and the second are resolved in two sorts of signs (Sebeok, 1994, pp. 12-13).

In our opinion and in accordance with Peirce who reformulated the classic notion of substitution in the medieval expression *aliquid stat pro aliquo* in terms of *interpretation, the sign is firstly an interpretant* (cf. Petrilli, 1998, I.1).

In fact, the Peircean terms of the sign include what may be called the *interpreted* sign on the side of the object, and the *interpretant* sign in a relation where it is the interpretant that makes the interpreted possible. The interpreted becomes a sign component because it receives an interpretation, but the interpretant is also a sign component endowed with potential for engendering a new sign. Therefore, where there is a sign, there are immediately two, and given that the interpretant can engender a new sign, there are immediately three, and so forth *ad infinitum* as conceived by Peirce with his notion of *infinite semiosis* or chain of deferrals from one interpretant to another.

To analyse the sign beginning from the object of interpretation, that is, the interpreted, means to begin from a secondary level. In other words, to begin from the object-interpreted means to begin from a point in the chain of deferrals, or semiosic chain, which cannot be considered as the starting point. Nor can the interpreted be privileged by way of abstraction at a theoretical level to explain the workings of sign processes. An example: a spot on the skin is a sign insofar as it may be interpreted as a symptom of sickness of the liver: this is already a secondary level in the interpretive process. At a primary level, retrospectively, the skin disorder is an interpretation enacted by the organism itself in relation to an anomaly which is disturbing it and to which it responds. The skin disorder is already in itself an interpretant response.

To say that the sign is firstly an interpretant means to say that the sign is firstly a response. We could also say that the sign is a reaction: but only on the condition that by *reaction* we mean "interpretation" (similar to Morris's behaviorism, but different from the mechanistic approach). The expression *solicitation-response* is preferable to *stimulus-reaction* in order to avoid superficial associations with the approaches they respectively recall. Even a "direct" response to a stimulus, or better solicitation, is never direct but "mediated" by an interpretation. Unless it is a *reflex action*, the formulation of a response involves identifying the solicitation, situating it in a context, and relating it to given behavioral parameters (whether a question of simple types of behavior, e.g., the prey-predator model, or more complex behaviors connected with cultural values, as in the human world).

The sign is firstly an interpretant, a response through which something else is considered as a sign and becomes its interpreted, on the one hand, and which is potentially able to engender an infinite chain of signs, on the other.

Consequently, the "ambiguity" of the concept of semiosis discussed in the entry "Semiosis" in *Encyclopedia of Semiotics*, edited by Paul Bouissac (1998), does not concern the term but the phenomenon of semiosis, at once a process and relation, activity and passivity, action *of* sign or action *on* sign, including sign solicitations and responses, interpreteds and interpretants.

In Peirce's view semiosis is a triadic process and relation whose components include sign (or representamen), object and interpretant. "A Sign, or Representamen, is a First which stands in such a genuine triadic relation to a Second, called its Object, as to be capable of determining a Third, called its Interpretant, to assume the same triadic relation to its Object in which it stands itself to the same Object" (CP 2.274). Therefore, the sign stands for something, its object, "not in all respects, but in reference to a sort of idea" (CP 2.228). However, a sign can only do this if it determines the interpretant which is "mediately determined by that object" (CP 8.343): semiosis is action of sign and action on sign, activity and passivity. "A sign mediates between the *interpretant* sign and its object" insofar as it refers to its object under a certain respect or idea, the ground, and determines the interpretant "in such a way as to bring the interpretant into a relation to the object, corresponding to its own relation to the object" (CP 8.332).

3.4. Centrality of the interpretant in the 'semiosic matrix'
Thure von Uexküll's model is so broad as to include sign processes from microsemiosis and endosemiosis to semiosis of higher organisms through to human biosemiotic meta-interpretation. It covers most of the complete catalogue of elements postulated for semiosis in the article entitled "Model of semiosis" by Martin Krampen (Posner, Robering, and Sebeok 1997-98, Vol. I, p. 248). This list includes the following 14 elements deemed necessary for a complete description of semiosis. Elements designated by a letter in parenthesis are located within the organism of the interpreter:

1. *the semiosis as a whole*: Z;
2. *the organism of the interpreter*: (O);
3. *the interpretandum* (signal): S;
4. *the channel*: Ch;

5. *the signifier* (the signal represented in the organism): (Rs);
6. *the interpretant*: (I);
7. *the signified* (the object represented in the organism): (Rg);
8. *the interpretatum* (object): G;
9. *the disposition for instrumental behavior*: (Rbg);
10. *the disposition for signaling behavior*: (Rsg);
11. *instrumental behavior*: (BG);
12. *signaling behavior*: (SG);
13. *external context*: (C);
14. *internal context*: (c).

On the basis of this list, a semiosis can be described in the following way:

> A semiosis Z is a process involving a channel Ch with an interpretandum S, which is related to an interpretandum G by being perceived and represented as a signifier (Rs) within the Organism (O) of its interpreter; the signifier (Rs) then being mediated by an interpretant (I) to connect with the signified (Rg), which represents the interpretatum G within (O). Via the interpretant (I), this process of symbolizing and referring triggers dispositions for instrumental behavior (rbg) and/or signaling behavior (Rsg); these are both related to the interpretatum G and terminate, via appropriate effectors, in overt instrumental behavior BG or signaling behavior SG, the latter supplying interpretanda for a further process of interpretation. Each semiosis Z is surrounded by other semioses and takes place in a context C external to (O) as well as a context (c) internal to (O). (Posner, Robering, and Sebeok 1997-98, Vol. I, p. 251)

This complex definition of semiosis is centered around the notion of *interpretant*. In fact, as we have already stated, the interpretant mediates between *solicitation* (interpretandum) and *response* (signaling behavior or instrumental behavior). In Peirce's view such mediation distinguishes a semiosis from a mere dynamical action — "or action of brute force" — which takes place between the terms forming a pair: on the contrary, semiosis results from a triadic relation: it "is an action, or influence, which is, or involves, a cooperation of three subjects, such as a sign, its object, and its interpretant" and it is not "in any way resolvable into action between pairs" (CP 5.484). The interpretant does not occur in physical phenomena or in non-biological interactions, in short, it does not occur in the inorganic world.

The definition of semiosis (quoted above) proposed by Krampen is illustrated graphically as a "semiosic matrix" (cf. Posner, Robering, and Sebeok 1997-98, Vol. I, p. 252, Fig. 5.1). A rhombus at the center of the *semiosic matrix* represents the interpretant I.

Most interesting is the pivotal role carried out by the interpretant in the semiosic matrix, indicated by placing the rhombus that represents the interpretant in the center.

3.5. The dialogic nature of sign and semiosis

The semiosic matrix which displays the various partial semiosic processes is used in the same article to illustrate graphically some other types of semioses such as *Pavlonian conditioning, the inference 'if ... then', hypothesis formation,* and a *'chain of thought'*. In all these types of semioses the semiosic matrix graph emphasizes the central role of the interpretant (cf. Posner, Robering, and Sebeok 1997-98, Vol. I, pp. 253-257).

Dialogue too is illustrated graphically through the semiosic matrix (cf. Posner, Robering, and Sebeok 1997-98, Vol. I, p. 260). The author of the article in question maintains that dialogue commences with signaling behavior from a sender intending to communicate something about an object. What is not taken into account by Krampen is that the 'if ... then' inference, hypothesis formation, and 'chain of thought' are dialogic forms in themselves.

In inference, in the hypothetical argument, and in the chain of interpreted and interpretant thought signs generally, dialogue is implied in the relation itself between the interpreted sign and the interpretant sign (cf. Ponzio, 1990, 1995, 1994).

The dialogic nature of sign in inference and the hypothetical argument has already been evidenced in previous writings (cf. Ponzio 1990: 197-214).

The degree of dialogism is minimal in deduction where the relation between the premises and the conclusion is *indexical*: here, once the premises are accepted the conclusion is obligatory.

In induction, it too is characterized by a unilinear inferential process, the conclusion is determined by habit and is of the *symbolic* order: identity and repetition dominate, though the relation between the premises and the conclusion is no longer obligatory.

By contrast, in abduction the relation between premises and conclusion is *iconic* and is dialogic in a substantial sense, in other words, it is characterized by high degrees of dialogism and inventiveness as well as by a high-risk margin for error. To claim that abductive argumentative procedures are risky is to say that they are mostly tentative and hypothetical with only a minimal margin for convention (symbolicity) and mechanical necessity (indexicality). Therefore, abductive inferential processes engender sign processes at the highest levels of otherness and dialogism.

The relation between sign (interpreted) and interpretant, as understood by Peirce, is a *dialogic* relation. We have already evidenced *the dialogic nature of the sign and semiosis*.

In *semiosis of information or signification* (Thure von Uexküll), where an inanimate environment acts as a "quasi-emitter" — or, in our terminology, where the *interpreted* becomes a *sign* only because it receives an interpretation by the interpretant which is a response — receiver interpretation is dialogic. Also, dialogue subsists in *semiosis of communication* (Thure von Uexküll) where the interpreted itself, before being interpreted as a sign by the interpretant is already an interpretant response calling for interpretation as a sign. However, dialogue also subsists in *semiosis of symptomatization* (Thure von Uexküll), where too the interpreted is an interpretant response (*symptom*) that similarly to the case of *semiosis of information or signification* does not arise for the sake of being interpreted as a sign.

Dialogue does not commence with signaling behavior from a sender intending to communicate something about an object. The whole semiosic process is dialogic, where the term 'dialogic' should be understood as *dia-logic*. The logic of semiosis as a whole and consequently the logic of Krampen's semiosic matrix is a *dia-logic*. The interpretant as such is "a disposition to respond," an expression used by Krampen to describe the dialogic interaction between a sender and a receiver (cf. Posner, Robering, and Sebeok 1997-98, I: 259).

Krampen's semiosic matrix in fact confirms the connection we have established between dialogue and semiosis. It shows that the two terms coincide, not only in the sense that dialogue is semiosis, but also in the sense that semiosis is dialogue — the latter being an aspect which would seem to escape Krampen. The dialogue process presented in the semiosic matrix is similar to the "if ... then" semiosic process, to hypothesis formation, chain of thought, and functional cycle after Jakob von Uexküll. In Krampen's article the semiosic matrix illustrates dialogue with two squares which represent the two partners, that is to say the sender and the receiver, where each has its own rhombus representing the interpretant. Despite this division, the graphic representation of dialogue is not different from the author's diagrams representing other types of semiosis. It could be the model, for example, of an "if...then" semiosis in which the two distinct interpretants are the premises and the conclusion of an argument in a single chain of thought.

3.6. Dialogue and the "functional cycle"

Jakob von Uexküll's "functional cycle" is a model for semiosic processes. As such it too has a dialogic structure and involves inferences of the "if...then" type which may even occur on a primitive level, as in Pavlovian semiosis or as prefigurations of the type of semiosis (where we have a "quasi-mind" interpreter) taking place during cognitive inference.

In the "functional cycle" the interpretandum produced by the "objective connecting structure" becomes an interpretatum and (represented in the organism by a signaling disposition) is translated by the interpretant into a behavioral disposition which triggers a behavior into the "connecting structure." Uexküll does not use a dialogic model. All the same, the point we wish to make is that in the "functional cycle" thus described a dialogic relation is established between an interpreted (interpretandum) and an interpretant (interpreted by another interpretant, and so forth), which does not limit itself to identifying the interpreted, but establishes an interactive relation with it.

Conversely, not only does the "functional cycle" have a dialogic structure, but dialogue in communication understood in a strict sense may also be analyzed in the light of the "functional cycle." In other words, the dialogic communicative relation between a sender who intends to communicate something about an object and a receiver may be considered, in turn, on the basis of the "functional cycle" model. The type of dialogue in question here corresponds to the processes described by the "functional cycle" as presented, in Thure von Uexküll's terminology, neither in *semiosis of information or signification*, nor in *semiosis of symptomatization*, but in *semiosis of communication*. Here the interpreted itself, before it is interpreted as a sign by the interpretant, is already an interpretant response addressed to somebody both to be identified and to receive the required *interpretant of answering comprehension*.

The entry "Dialogue" is lacking in the *Handbook of Semiotics* by Winfried Nöth (1990). However this term is listed in the "Index of subjects and terms" which informs us that the subject is treated in a chapter entitled "Communication and semiosis" (Part 3). Here the "functional cycle" is also mentioned (Nöth, 1990, pp. 176-180). This indicates the implications of Uexküll's biosemiosic "functional cycle" for the problem of the relation between dialogue and communication. Different communication models are discussed showing how biological models, which describe communication as a self-referential autopoietic and semiotically closed system (such as the models proposed by Maturana, Varela, and Thure von Uexküll), are radically opposed to both the linear (Shannon & Weaver) and the circular (Saussure) paradigms. As reported by Nöth (1990, p. 180), Thure von Uexküll (1981, p. 14) demonstrates that Jakob von Uexküll's (1982, p. 8) biosemiosic functional cycle has this feature of autonomous closure and therefore reacts to its environment only according to its internal needs.

The theory of autopoietic systems is incompatible with dialogism only if one subscribes to a trivial conception of dialogue based on a communication model that describes communication as a linear causal process. This is a process moving from source to destination. Similarly there is incompatibility between autopoietic systems and dialogism, if dialogue is conceived as based on the conversation model governed by the turning around together rule. Also, the autopoietic system calls for a new notion of creativity. Furthermore, there remains the question of how the principle of autonomous closure be compatible with dialogue conceived as the inner structure of the individual, therefore with creativity and learning. As Maturana (1978, pp. 54-55) would seem to suggest, it is possible to conceive dialogic exchange as opposed to communication understood as a linear process from source to destination or as a circular process in which participants take turns in playing the part of sender and receiver: This dialogue, says Maturana, should be conceived as "pre- or anticommunicative interaction."

3.7. Dialogism and biosemiosis
Concerning the Bakhtinian notion of "dialogism" we have observed (see Petrilli & Ponzio, Semiotics Unbounded, 2002, Part One, III. 1.4) how in Bakhtin's view dialogue does not consist in the communication of messages, nor is it an initiative taken by self. On the contrary, the self is always in dialogue with the other, that is to say, with the world and with others, whether it knows it or not; the self is always in dialogue with the word of the other. Identity is dialogic. Dialogism is at the very heart of the self. The self, "the semiotic self" (see Sebeok, Petrilli, & Ponzio, 2001), is dialogic in the sense of a species-specifically modeled involvement with the world and with others. Self is implied dialogically in otherness, just as the "grotesque body" (Bakhtin, 1965) is implied in the body of other living beings. In fact, in a Bakhtinian perspective *dialogue* and *intercorporeity* are closely interconnected: there cannot be

dialogue among disembodied minds, nor can dialogism be understood separately from the biosemiotic conception of sign.

As we have already observed, Bakhtin's main interpreters such as Holquist, Todorov, Krysinsky and Wellek have all fundamentally misunderstood Bakhtin and his concept of dialogue. This is confirmed by their interpretation of Bakhtinian dialogue as being similar to dialogue theorized by such authors as Plato, Buber, and Mukarovsky.

For Bakhtin dialogue is the embodied, intercorporeal, expression of the involvement of one's body, which is only illusorily an individual, separate and autonomous body, with the body of the other. The image that most adequately expresses this idea is that of the "grotesque body" (cf. Bakhtin, 1965) in popular culture, in vulgar language of the public place, and above all in the masks of carnival. This is the body *in its vital and indissoluble relation to the world and to the body of others*. With the shift in focus from identity (whether individual, as in the case of consciousness or self, or collective, that is to say, a community, historical language, or a cultural system at large) to alterity, a sort of *Copernican revolution* is accomplished. Bakhtinian critique conducted in terms of dialogic reason not only interrogates the general orientation of Western philosophy, but also the dominant cultural tendencies that engender it.

The "Copernican revolution" operated by Bakhtin in relation to the conception of self, identity, and consciousness involves all living beings and not just the human. Consciousness implies a dialogic relation which includes a witness and a judge. This dialogic relation is not only present in the strictly human world but also in the biological. Says Bakhtin

> When consciousness appeared in the world (in existence) and, perhaps, when biological life appeared (perhaps not only animals, but trees and grass also witness and judge), the world (existence) changed radically. A stone is still stony and the sun still sunny, but the event of existence as a whole (unfinalized) becomes completely different because a new and major character in this event appears for the first time on the scene of earthly existence — the witness and the judge. And the sun, while remaining physically the same, has changed because it has begun to be cognized by the witness and the judge. It has stopped simply being and has started being in itself and for itself ... as Well as for the other, because it has been reflected in the consciousness of the other ... . ('From notes made in 1970-71', in Bakhtin, 1986, p. 137)

3.8. The biological basis of Bakhtinian dialogue and "great experience"
At this point I wish to evidence a possible connection between Sebeok's biosemiotic conception and Bakhtin's dialogic conception. It would seem that these two authors are very distant from each other. In reality this is not true. We also find an interest in biology in the formation process of Bakhtin's thought. Bakhtin develops his conception of dialogue in close connection to the biological studies of his time and according to a totalizing view expressed in the concept of biosphere as proposed by Vernadsky. In both Sebeok's and Bakhtin's conception all living beings on the planet Earth are, directly or indirectly, closely interrelated and interdependent in spite of their apparent autonomy and separation.

Bakhtinian dialogue is not the result of an attitude taken by the subject toward the other, but rather it is the expression of the living being's biosemiosic impossibility of closure and indifference toward its environment with which it constitutes a whole system, called by Bakhtin *architectonics*. In human beings architectonics becomes an 'architectonics of answerability', a semiotic consciousness of 'being-in-the-world-without-alibis'. It may be limited to a small sphere — that is to say, the restricted life environment of a single individual, one's family, professional, work, ethnic, religious group, culture, contemporaneity — or, on the contrary, it may be extended, as consciousness of the 'global semiotic' order (the term is Sebeok's), to the whole world in a planetary or solar or even cosmic dimension (as auspicated by Victoria Welby). Bakhtin distinguishes between 'small experience' and 'great experience'. The former is narrow-minded experience. Instead

> in the great experience, the world does not coincide with itself (it is not what it is), it is not closed and finalized. In it there is memory which flows and fades away into the human depths of matter and of boundless life, experience of worlds and atoms. And for such memory the history of the single individual begins long before its cognitive acts (its cognizable 'Self').. (Bakhtin's 'Notes of 1950', in Bakhtin, 1996, p. 99)

We must not forget that Bakhtin authored an article in 1926 entitled "Contemporary vitalism," in which he discusses biological and philosophical problems. This article was signed by the biologist Ivan Ivanovich Kanaev (see Kanaev, 1926) and is an important tessera for the reconstruction of Bakhtin's thought from the time of his early studies. Similar to the development of research by the biologist Jakob von Uexküll, in Bakhtin we also find an early interest in biology specifically in relation to the study of signs.

As anticipated (Part One, III.1.5.), this article on vitalism was written during a period of frenzied activity for Bakhtin, during the years 1924-29, in Petersburg, then Leningrad.

In this productive period of his life he published four books on different subjects (Freud, Russian Formalism, philosophy of language, Dostoevsky's novel), only the last of which is under his name, while the others (together with several articles) are signed Voloshinov or Medvedev.

In Petersburg Bakhtin lived in Kanaev's apartment for several years, and Kanaev contributed to Bakhtin's interest in biology as well as to the influence exerted by the physiologist Ukhtomsky on his conception of the "chronotope" in the novel. Jakob von Uexküll is also named in Bakhtin's text on vitalism.

In "Contemporary Vitalism" Bakhtin's criticism of vitalism — the conception that theorizes a special extra-material force in living beings as the basis of life processes — is directed above all against the biologist Hans Driesch who interpreted the organism's homeostasis in terms of full autonomy from its surrounding environment. On the contrary, Bakhtin opposes the dualism of life force and physical-chemical processes in his own description of the interaction between organism and environment, maintaining that the organism forms a monistic unit with the surrounding world. The relation of body and world is a dialogic relation in which the body answers to its environment modeling its world.

3.9. Rabelais's world as the world's biosemiotic consciousness

The category of the "carnivalesque" as formulated by Bakhtin and the role he assigns to it in his study on Rabelais can only be adequately understood in the light of his global (his 'great experience') and biosemiotic view of the complex and intricate life of signs.

The original title of Bakhtin's book on Rabelais, literally *The Work of François Rabelais and Popular Culture of the Middle Ages and Renaissance*, stresses the intricate connection between Rabelais's work, on the one hand, and the view of the world as elaborated by popular culture (its ideology, its *Weltanschauung*) as it evolves from Ancient Greek and Roman civilization to the Middle Ages and Renaissance, on the other, which in Western Europe is followed by the significant transition into bourgeois society and its ideology.

Bourgeois ideology conceives bodies as separate and reciprocally indifferent entities. Thus understood, bodies only have two things in common: firstly, they are all evaluated according to the same criterion, that is to say, their capacity for work; secondly, they are all interested in the circulation of goods, work included, to the end of satisfying the needs of the individual. Such ideology continued into Stalinist Russia which coincides with the time of Bakhtin's writing, and into the whole period of real socialism where only work and the capacity for production were considered as community factors, in other words, work and productivity were the sole elements linking individuals to each other. Therefore, beyond this minimal common denominator individual bodies remain reciprocally indifferent to each other and separate.

The carnivalesque participates in the "great experience" which offers a global view of the complex and intricate life of bodies and signs. The Bakhtinian conception emphasizes the inevitability of vital bodily contact showing how the life of each one of us is implicated in the life of others. In what may be described as a "religious" (from Latin *religo*) perspective of the existent, therefore, this conception underlines the bond uniting all living beings to each other. Furthermore, the condition of excess is emphasized, of bodily excess with respect to a specific function and sign excess with respect to a specific meaning: signs and bodies — bodies as signs of life — are ends in themselves. On the contrary, the minor and more recent ideological tradition is vitiated by reductive binarism which sets the individual against the social, the biological against the cultural, the spirit against the body, physical-chemical forces against life forces, the comic against the serious, death against life, high against low, the official against the non-official, public against private, work against art, work against non official festivity. Through Rabelais Bakhtin recovered the major tradition and criticized the minor and more recent conception of the individual body and life inherent in capitalism as well as in real socialism and its metamorphoses. Dostoevsky's polyphonic novel was in line with the major tradition in

Weltanschauung, as demonstrated by Bakhtin in the second edition (1963) of his 1929 book.

Self cannot exist without memory, and structural to both the individual memory and social memory is otherness. In fact, the kind of memory we are alluding to is the memory of the immediate biosemiotic "great experience" (in space and time) of indissoluble relations to others lived by the human body. These relations are represented in ancient forms of culture as well as in carnivalized arts: however, the sense of the "great experience" is anesthetized in the "small," narrow-minded, reductive experience of our time.

To conclude: modeling and dialogism are pivotal concepts in the study of semiosis. Communication is only one kind of semiosis that — together with the semiosis of information or signification and the semiosis of symptomatization — presupposes the semiosis of modeling and dialogism. This emerges clearly if in accordance with Peirce and his reformulation of the classic notion of *substitution* in terms of *interpretation*, we consider the sign first of all as an interpretant, that is to say as a dialogic response foreseen by a specific type of modeling.

4. Sebeok's Semiotics and Education

4.1. The role of signs in the educational process

In his programmatic theses on "Semiotic and the School," Morris (1946) observed that to use semiotics as the foundation for education does not mean to introduce semiotics as a separate discipline and a technical terminology into the early levels of the school system, but to show teachers, at every stage of the educational process, the role played by signs in our performances, how they serve various ends, their adequacy or inadequacy in actual communication.

> At the level of higher education, a specific and detailed study of semiotic can serve to raise to fuller awareness the training in the adequate use of signs which should have occurred throughout the earlier levels. (Morris 1946, in 1971: 326)

In other words, the contribution of semiotics to education is methodological, ranging from the theoretical foundations of education to the particular aspects of teaching and learning. Semiotic research in education is not only an area of education but also of semiotics. In his *Handbook of Semiotics*, Winfried Nöth (1990) dedicates a paragraph ("Teaching") of chap. III, "Semiosis, Code, and the Semiotic Field" to the relations between semiotics and education. These include: issues of semiotics in teaching which is concerned with educational interactions as processes of semiosis and communication (cf. pp. 221-222); the role of semiotics in the teaching of school subjects (native language teaching, foreign language teaching; nonverbal and visual communication in the foreign culture; the semiotics of culture in foreign language teaching; semiotic foundations of teaching methodology; visual arts and media languages as school subjects (cf. pp. 222-223); finally semiotics as an explicit subject of teaching in University programs and in schools (cf. pp. 223-224).

Nöth includes Sebeok among those who have contributed both to the description of teaching programs and syllabi for semiotics as a mayor or minor in university studies (see Sebeok, 1976, pp. 176-180; 1979, pp. 272-279), and to the semiotic foundations of the theoretical and practical aspects of education (see Sebeok, Lamb, & Regan, 1988).

But Sebeok's greatest semiotic contribution to methodological studies in the field of education comes from his innovative ideas in the field of semiotics. The purpose of *The Body in the Sign. Thomas A. Sebeok and Semiotics* by Marcel Danesi (1998) is precisely to consider the implications of Sebeok's work for education in a general sense. This book explains the Sebeokean approach to semiotics in order to introduce Sebeok's semiotics to educators and pedagogical researchers and to show what his work may have to offer to their field of research.

4.2. Implications of Sebeok's work for education
As we have said, Sebeok's approach to semiotics — which may be called "global semiotics" or "semiotics of life" — is the one that, at present, most assumes an interdisciplinary focus in research. Consequently the contribution to educational practices and to the future of research in the educational domain does not consist in constructing an all-embracing theory of learning. The semiotic perspective on the use of verbal and nonverbal signs in the learning process does not exclude other disciplinary views.

Sebeok extends the boundaries of traditional semiotics which is vitiated by a fundamental error, that of mistaking a part (that is, human signs and in particular verbal signs), for the whole (that is, all possible signs, human and nonhuman). Instead, Sebeok's "global semiotics," as described above, is the place where the "life sciences" and the "sign sciences" converge, therefore the place of consciousness of the fact that the human being is a sign in a universe of signs. Such an approach presupposes a critique of anthropocentrism and of glottocentrism which indubitably has positive effects when a question of developing educational aims and methods.

According to Danesi (1998, p. 61), the principal suggestions that may come from Sebeokean semiotics for educational research and practice are: first and foremost that learning and semiosis are co-occurring and complementary functions; and second that all types of learning in human development, as results in childhood, are modeling processes which may be described as a "flow" from iconicity to cultural symbolism. Children model knowledge and skill acquired through the body first iconically and then symbolically as they adapt human natural primary modeling to the forms of secondary and tertiary modeling of their cultural context. Sebeok's typology of human modeling is pivotal in a semiotics of education.

As Danesi (1998, p. 28) explains referring to the second chapter of Sebeok (1986), it is a mistake to think of language has having developed primarily out of a need to communicate. We must distinguish with Sebeok between *language* and *speech*. Language is essentially "mind work," speech is "ear and mouth work."

Thanks to syntax, human language, like Lego building blocks, can reassemble a limited number of construction pieces in an infinite number of different ways. As a modeling device language can produce an indefinite number of models, which brings us back to the "play of musement." The plurality of languages and "linguistic creativity" (Chomsky) testify to the capacity of language, understood as a primary modeling device, for producing numerous possible worlds.

In spite of insistence on the "creative character of (verbal) language," Chomsky's linguistics is unable to explain the plurality of natural languages (and "inner plurilingualism" in any natural language). The reason is that Chomskyian linguistics presupposes an innate Universal Grammar. On the contrary, the fact that human beings have invented numerous natural languages is the direct result of the modeling capacity, and therefore of the capacity to invent multiple worlds. In other words, the plurality of languages derives from the propensity of language for the "play of musement" or, as Giambattista Vico says in "poetic logic," characteristic of human beings.

4.3. Education to mutual adjustment of language and speech

Concerning the relation between language and speech, Sebeok remarks that it required a plausible mutual adjustment of the encoding with the decoding capacity. On the one hand, language was "exapted" for communication (first in the form of speech, i.e., for "ear and mouth work," and later in the form of script, and so forth), and, on the other, speech was exapted for (secondary) modeling, i.e. for "mind work." "But," Sebeok adds, "since absolute mutual comprehension remains a distant goal, the system continues to be fine-tuned and tinkered with still" (1986, p. 56).

It is opportune to draw attention to important implications of this comment for education and pedagogical research, in addition to Danesi's argument on the relevance of Sebeok's research to the educational domain. The significance of the following observations for teaching and learning is unquestionable:

> It is reasonable to suppose that the adjustment, or fine-turning, of the encoding capacity required by speaking to the decoding capacity required to understand speech, and vice versa, took about two million years to achieve at least partially. (Full understanding is a rare commodity; most of the time most of us don't quite grasp what another human being is trying to tell us.) Even today, humans have no special organ for speech, which is formed by a tract originally designed for two entirely different biological functions: the alimentary and the respiratory. Speech is then received, as any other sound, by the ear, which has still another phylogenetic source and is a rather newly acquired sensory receptor. (Sebeok, 1986, p. 70)

In fact, this very discrepancy between thought and speech in communication may be corrected only through the educational process. Problems of speech are also problems of thought, and vice versa. They are also problems of communication. Sebeok's reflections on the origins of language and speech can help to eliminate misunderstandings in communication both on the part of expression and on the part of comprehension.

Another concept of Sebeok's semiotics has consequences for the theory of language and therefore for the linguistic educational process. The human mind moves among meanings and concepts in a way that Danesi (1995, 2000) calls "imaginative mental navigating" into a web of interpretive courses according to associations included in the complex system or macroweb that we usually call "culture." It follows that both the notions of "linguistic competence" and "communicative competence" are inadequate and insufficient to explain the behavior of thinking and speaking, i.e. the capacity for verbalization and reasoning. Both these competencies are part of an organic *conceptual competence*, the ability, as Danesi (1995, 2000) shows, to convert mental schemes of various conceptual provenance into linguistic and communicative structures and consequently to create messages which are conceptually and culturally appropriate and pertinent. This conceptual competence consists in three sub-competencies: a) *metaphoric competence*, that is, the ability to produce metaphorically appropriate concepts; b) *reflexive competence*, that is, the ability to select linguistic structures and categories that reflect appropriate conceptual realms of messages; and c) *cultural competence*, that is, the ability to sail across different discourse fields and conceptual realms of which the message makes use.

Real "linguistic creativity" consists of the capacity to form new metaphoric associations, to propose new knowledge combinations, and to invent new representations. This is the basis of human symbolic behavior, that is to say, of species-specific human primary, secondary and tertiary modeling systems.

Referring to the associative character of verbal language and thought, with Danesi and in opposition to the Cartesian model of the thinking subject, we may say that human beings are anything but rational thinkers; they are rather ingenious "guessers." "To guess," as says Peirce, is the characteristic of argument, and the greater the risk in associating different terms that belong to distant fields in the macroweb of culture, the stronger is the argument to invent and innovate.

4.4. Semiotics and foresight of "proximal development"

Attempting to sum up what Sebeok has taught a whole generation of semioticians in one phrase, Danesi chooses the following: "the body is in the sign," i.e. life is defined by semiosis. In the human species this means that, according to Sebeok, semiosis is the bond that links together body, mind, and culture (see Danesi, 1998, p. 16). *The Body in the Sign* is the title of Danesi's book on Sebeok and semiotics. Sebeok examines the manifest patterns of semiosis in nature and culture showing persuasively that in anthroposemiosis there exists an inextricable nexus among sign, body and culture.

In the dialogue "Semiotics in Education" with Lamb and Regan, Sebeok observes:

> [Peirce said] that all this universe is perfused with signs. Then he added a thoughtful statement that it may indeed be composed exclusively of signs. The difficulty with that statement is that it is not verifiable. If you believe that the universe is perfused with signs, and if you believe, as Professor Lamb said, that we all have a mental model of the universe, an internalized mental model that admits into the mind nothing but signs, then, if there is any thing else left, it is not verifiable and therefore

not knowable. This is known as the radical idealistic position. As a mild idealistic position, we would say that we sense that maybe there is something out there. For example, Heraclitus said there is something out there that he called the "logos." But who knows? And I think that this radical idealistic position is in conformity with some versions of quantum mechanics. So on this point I am not sure whether we agree or disagree. If we disagree, it is up to him to prove that there is something out there that I cannot get at by means other than through signs. (Sebeok, in Sebeok, Lamb, and Regan 1988, p. 12)

Sebeok also argues that iconicity is the default form of semiosis, documenting in vastly different species the manifestation of the capacity to produce signs that stand in some direct simulative relation to their referents. In other words, iconicity is, according to Sebeok, a basic signifying strategy in various life forms. The iconic mode of representation is the relation of the sign with its referent through replication, simulation, imitation, or resemblance. In his works (1979, 1986, 1991, 1994), Sebeok shows the variety of manifestations of iconicity in different species. Iconic signs can thus be vocal, visual, olfactory, gustatory, or tactile in their form. It may be that in humans too all signs start out as a simulative relation to their referential domains. Like Peirce, Sebeok sees iconicity as the primordial representational strategy in the human species. Danesi (cf. 1998, p. 10) personally considers iconicity as an aspect of utmost relevance in the study of signs. He emphasizes the important role of iconicity — documented by Sebeok in the final three chapters of his book of 1986 — in the bond that links semiosis, body, mind, and culture. This inextricable nexus manifests itself in the form of iconical representational behavior. "Iconicity is, in effect, evidence of this nexus" (Danesi, 1998, p. 37). Danesi refers to the conception that the iconic mode of representation is the primary means of bodily semiosis as the *iconicity hypothesis* (Danesi, 1998, pp. 18-20).

Consequently another principle of Sebeokian semiotics is "the sense-implication hypothesis" (Danesi, 1998, p. 17), which suggests that semiosis is grounded in the experiential realm of sense. This principle has a philosophical antecedent in John Locke — according to whom all ideas came from sensation first and reflection later — but it is connected with the modeling theory: what is acquired through the body is differently modeled through the innate modeling system possessed by different species. In fact, a species perceives according to its own particular anatomical structure and kind of modeling system. Thanks to its species-specific modeling system, called by Sebeok language, *Homo* is not only a sophisticated modeler of the world, but also has a remarkable ability to recreate his world in an infinite number of different forms.

What in human modeling, as described by Sebeok's biosemiotics, is relevant for education is what Danesi calls the *natural learning flow*, that is to say, the semiosic process in which children acquire knowledge: this process takes place through the body and human primary modeling system and proceeds from iconicity to the forms of modeling that children learn in the cultural context The *natural learning flow principle* implies, says Danesi (Danesi, 1998, p. 61), that the semiosic capacities of the learner and the determination of his semiosic stage — rather than the subject matter to be learned — should therefore be the focus of education. The main implication of Sebeok's modeling theory for education is of a *methodological nature*.

If the teacher is familiar with the forms of the semiosic process in human learning, he/she would be in a better position to help the learner acquire knowledge and skill more effectively and efficiently. In fact the key to successful learning, says Danesi, lies, arguably, in determining at what point the learning phase is ready to be overtaken by the following, i.e., what the Russian psychologist Vygotsky (1934), to whom Danesi explicitly refers, calls the "proximal zone" of learning. The semiotic approach to education, as the psychologist but also semiotician Vygotsky claimed, is indispensable for an appropriate understanding of the "zones of proximal development" of each particular learner.

4.5. Global semiotics and education to responsibility for life
We wish to add another argument to those proposed by Danesi on the implications of Sebeok's global semiotics for education. As goals in education we count the capacity for criticism, social awareness, and responsible behavior. Sebeok's semiotics of life may be introduced to an audience of educators and pedagogical researchers thanks to its implications for an adequate consciousness and comprehensive interpretation of communication under present day conditions, i.e. in the phase named 'globalization'.

The present age is characterized by the automated industrial revolution, by the global market and therefore by the pervasiveness of communication in the whole production cycle (production, exchange, consumption). In this context, communication is exploited for capitalistic profit. This is a danger for communication and includes the risk of destroying communication itself understood as the possibility of life over the whole planet Earth. (cf. Ponzio & Petrilli, 2000).

We may gain consciousness of the risk that we are running and explain it to others, particularly the new generations — and teach them responsibility — by assuming a point of view that is as global as our social system; this implies the capability of grasping the link between communication and life. This point of view is offered to us by Sebeok's semiotics.

Sebeok's planetary perspective in both a spatial and temporal sense will permit the necessary distance and indeclinable responsibility (a responsibility without alibis) for an approach to contemporaneity that does not remain imprisoned within the boundaries of contemporaneity itself (see Petrilli, introduction to Italian translation of Sebeok, 1991; see also Petrilli, 1998; Ponzio & Petrilli, 2002).

The interdisciplinary focus of Sebeok's global semiotics and attention to the signs of the interconnection between body and species are the presuppositions of an approach to education which is free from stereotypical stereotypical, limited, and distorted ideas, and from the practices of communication under present day conditions. This is another implication of Sebeok's work for education and another possible meaning of the phrase chosen by Danesi to sum up what Sebeok said: "The body is in the sign," i.e., semiosis is the bond that links body, mind and culture.

References

Bakhtin, M. (1963). *Problemy tvorchestva Dostoevskogo* (2nd ed). Leningrad: Priboj.
Bakhtin, M. (1965). *Rabelais and His World* (K. Pomorska, Ed. & Trans.). Cambridge: The MIT Press.
Bakhtin, M. (1986) *Speech Genres & Other Late Essays* (C. Emerson and M. Holquist, Eds., V. W. McGee, Trans.). Austin, TX: University of Texas Press.
Bakhtin, M. (1996). *Sobranie Sochinenij* (Collected Papers, vol. V.) Moscow: Russkie Slovari.
Bouissac, P. (Ed.). (1998). *Encyclopedia of Semiotics.* New York: Oxford University Press.
Bouissac, P., Herzfeld, M., & Posner, R. (1986). *Iconicity: Essays on nature and culture. Festschift for Thomas A. Sebeok on the his 65th birthday.* Tübingen: Stauffenburg.
Danesi, M. (1995). *Giambattista Vico and the cognitive science enterprise.* New York: Peter Lang.
Danesi, M. (1998). *The body in the sign: Thomas A. Sebeok and semiotics.* Toronto: Legas.
Danesi, M. (2000). *Metafora, lingua concetto. Vico e la linguistica cognitiva.* Bari: Edizioni dal Sud. (Intro. by Augusto Ponzio)
Lucid, D. P. (Ed.) (1977). *Soviet semiotics: An anthology.* Baltimore: Johns Hopkins University Press.
Maturana, H. R. (1978). Biology of language: The epistemological reality. In G.A. Miller & E. Lenneberg(Eds.), *Psycologyand biology of language and thought* (pp. 27-63). New York: Academic Press.
Morris, C. (1971). *Writings on the general theory of signs* (T. A. Sebeok, Ed.). The Hague-Paris: Mouton.
Nöth, W. (1990). *Handbook of semiotics.* Bloomington, IN: Indiana University Press.
Ogden, C. K. (1994). *C K. Ogden and linguistics* (5 vols. T. W. Gordon, Ed.). London: Routledge-Thoemmes Press.
Peirce, C. S. (1931-1966). *Collected Papers of Charles Sanders Peirce* (8 vols., C. Hartshorne, P. Weiss, & A. W. Burks, Eds.). Cambridge, MA: Harvard University Press. (Our references are to *CP*, volume#.paragraph #)
Petrilli, S. (1998). *Teoria dei segni e del linguaggio.* Bari: Graphis.
Petrilli, S. & Ponzio, A. (2002). *Semiotics unbounded: Interpretive routes through the open network of signs.* (Unpublished manuscript under review.)
Ponzio, A. (1990). *Man as a sign: Essays on the philosophy of language* (S. Petrilli, Ed. & Trans.). Berlin: Mouton de Gruyter. (Appendix I & II by S. Petrilli)
Ponzio, A. (1994). *Fondamenti di filosofia del linguaggio* (with P. Calefato and S. Petrilli). Rome-Bari: Laterza.
Ponzio, A. (1995). *Segni per parlare dei segni. Signs to talk about signs* (bilingual text, S. Petrilli, Eng. Trans.). Bari: Adriatica.
Ponzio, A. & Petrilli, S. (2000). *Il sentire nella comunicazione globale.* Rome: Meltemi.
Ponzio, A. & Petrilli, S. (2003). *Semioetica.* Rome: Meltemi.
Posner, R., Robering, K., & Sebeok, T. A., (Eds.). (1997-98). *Semiotik semiotics: A handbook on the sign-theoretic foundations of nature and culture* (3 vols.). Berlin: Walter de Gruyter, 1998 (vol. 3 is forthcoming).
Rudy, S. (1986). Semiotics in the USSR. In T A. Sebeok and J. Umiker-Sebeok (Eds.), *The semiotic sphere*, Chpt. 25. New York: Plenum Press.
Sebeok, T. A. (1976). *Contributions to the Doctrine of Signs* (2nd ed). Bloomington, IN: Indiana University Press. (It. trans. by M. Pesaresi. Contributi alla dottrina dei segni. Milano: Feltrinelli, 1979.)
Sebeok, T. A. (1979). *The Sign & Its Masters.* Austin, TX: The University of Texas Press. (It. trans. and intro. by S. Petrilli, Il segno e i suoi maestri. Ed. intro. and trans. by S. Petrilli. Bari: Adriatica, 1985.)
Sebeok, T. A. (1981). *The Play of Musement.* Bloomington, IN: Indiana University Press. (It. trans. by M. Pesaresi, Il gioco del fantasticare. Milan: Spirali, 1984).
Sebeok, T. A. (1986). *I think I am a verb.* New York-London: Plenum Press. (It. trans. and intro. by S. Petrilli, *Penso di essere un verbo.* Palermo: Sellerio, 1990.)
Sebeok, T. A. (1991). *A sign is just a sign.* Bloomington, IN: Indiana University Press. (It.trans. and intro. by S. Petrilli, *A sign is just a sign. La semiotica globale.* Milan: Spirali, 1998.)
Sebeok, T. A. (1994). *Signs. An Introduction to Semiotics.* Toronto: Toronto University Press.
Sebeok, T. A., Lamb, S. M., & Regan, J. O. (1988). *Semiotics in education: A dialogue.* Claremont, CA: Claremont Graduate School.
Sebeok, T. A. & Danesi, M. (2000). *The forms of meanings: Modeling systems theory and semiotic analysis.* Berlin: Mouton de Gruyer.
Sebeok, T. A., Petrilli, S., & Ponzio, A. (2001). *Semiotica dell'io.* Rome: Meltemi.
Vygotsky, L. S. (1962). *Thought and language.* Cambridge: The MIT Press. (Original work published in 1934)

Thomas Sebeok: Mister (Bio)semiotics
An obituary for Thomas A. Sebeok

Søren Brier

Thomas A. Sebeok was born on November 9, 1920, in Budapest, Hungary, and died peacefully in his home in Bloomington, Indiana, on December 21, 2001. He was a pioneer in semiotics and the creator of the field of biosemiotics. He belongs to the most renowned exponents of semiotics in the second half of the 20th century. By means of his scientific, institutional, and editorial efforts he exerted steady influence on the development of semiotics as a transdisciplinary field of inquiry; the doctrine of semiotics he called it. He produced numerous books and even more essays and other writings on general semiotics, zoösemiotics and biosemiotics, as well as linguistics, psycholinguistics, mythology, folklore, ethology, stylistics, and theory of art.

In 1991 he was awarded the title of distinguished professor emeritus of linguistics, semiotics, anthropology and Central Asian Studies at Indiana University, Bloomington. When the International Association for Semiotic Studies (IASS-AIS) was founded in Paris on January 21, 1969, Sebeok was elected editor-in-chief of the newly created journal Semiotica, and he fulfilled this duty with unflagging devotion until the end of his life. In this function he was also a member and important promoter of the Board of the (IASS-AIS) since its foundation.

For decades he was Series Editor, responsible for a number of leading book series *Advances in Semiotics*, *Approaches to Semiotics*, *Approaches to Applied Semiotics*, and *Topics in Contemporary Semiotics*, and he was the General Editor of the standard reference work *Encyclopedic Dictionary of Semiotics* (1986; recently published in a revised and enlarged version.)

Sebeok also served as chairman of the *Indiana University Research Center for Language and Semiotic Studies*, was a professor of anthropology and of Uralic and Altaic Studies and a fellow of the Folklore Institute. In the semiotic study of human communication, for instance, he would examine not only spoken conversation but also nonverbal paralinguistic signs such as facial expressions and body movements which carry information along with the spoken words – sometimes in a manner that contradictory the words.

Sebeok first left Hungary in 1936 to study at Magdalene College, Cambridge University. The following year, he immigrated to the United States and obtained citizenship in 1944. He earned a bachelor's degree at the University of Chicago in 1941 and an M.A. (1943) and Ph.D. (1945) at Princeton University. While he was at Princeton, he also commuted to Columbia University to continue his study of linguistics under Roman Jakobson, who was his dissertation director. At the same time

Sebeok was also a disciple of Charles W. Morris in Chicago who with Jakobson was among the most prominent progenitors of the transdisciplinary view of semiotics

In 1943 Sebeok went to Indiana University to assist in running the largest Army Specialized Training Program in foreign languages, and after a while he became director of that program. Soon after this he created the renowned Department of Uralic and Altaic Studies there and was offered the directorship of the *Research Center for Anthropology, Folklore, and Linguistics*. It was the transformation of this center into the *Research Center for Language and Semiotic Studies* (RCLSS) in 1956 that finally generated the famous *Center for Semiotics*. For decades this was one of the most influential academic institutions in the world of semiotics.

As a linguist studying Finno-Ugric languages, Sebeok's fieldwork took him to Central and Eastern Europe, including Lapland and the former Soviet Union. He also carried out linguistic studies in the former Mongolian People's Republic, Mexico and in the U.S. (among the Winnebago Indians of Wisconsin and the Laguna Indians of New Mexico). In addition to these studies in grammar and phonology, his interest in anthropology, folklore and literary studies led to publications dealing with folksongs, charms, games, poems and the supernatural.

He published a ground-breaking volume on *Myth* in 1955, and in 1960, *Style in Language*. At the same time, he contributed to the creation of the new field of psycholinguistics, publishing with Charles Osgood, the famous classic text, *Psycholinguistics,* in 1954. He also made some of the first computer analyses of verbal texts.

By 1960, Sebeok had established himself as a scholar known for crossing academic boundaries not only in his own research, but also in collaboration with scholars in adjacent fields. This carried over into his roles as an editor of books and journals, a founder and officer of several academic organizations, conference organizer, and mentor.

Several fellowships at the *Stanford University Center for Advanced Studies in the Behavioral Sciences* gave him the opportunity to return to his avocation, biology. In the 1960s he turned to the study of human nonverbal and animal communication, publishing several seminal volumes on these topics that were important contributions to the comparative study of communication and signification. The publication of his classic *Approaches to Semiotics* in 1964 marked the beginning of his development of a general semiotics that extends beyond human language.

The transmission of information among animals was also the subject of the book *Animal Communication* (1968), which he edited. His research succeeded in broadening the definition of semiotics beyond human language and culture to encompass human nonverbal communication as well as communication between and sign processes within all living organisms. Sebeok's name is associated most of all with the term zoösemiotics, the study of animal sign use. It was coined in 1963 and deals with species-specific communication systems, their signifying behavior. Zoösemiotics is concerned more with a synchronic perspective than ethology, which focuses primarily on the diachronic dimension.

Later Sebeok decided that zoösemiotics rests on a more comprehensive science of biosemiotics. This global conception of semiotics – namely biosemiotics– equates life with sign interpretation or mediation. He was most proud of having inspired a group of theoretical biologists and semioticians to pursue this field of investigation.

Although biosemiotics is already prefigured in Jakob von Uexküll's *Umweltlehre* Sebeok fruitfully combined the influences of von Uexküll and Charles S. Peirce, to merge them into an original whole, in an evolutionary perspective, arriving at the thesis that symbiosis and semiosis are one and the same. Biosemiotics finds its place as a master-science which encompasses the parallel disciplines of ethology and developmental psychology. As Uexküll was one of Konrad Lorenz' most important teachers, it is no wonder Uexküll's research program was almost identical with that of ethology except for his lack of belief in evolution.

The empiricist and natural science readings Sebeok offers for communication were new to the semiotic field. References to animal models are made throughout his work in the context of ethology. The approaches of ethology and sociobiology have been controversial and, in their applicability to human culture and society, accused of reductionism. Sebeok shows that some of this controversy may find itself played out in the new transdisciplinary framework of biosemiotics. In 1992 he and his wife Jean Umiker-Sebeok published *The Semiotic Web 1991* as a volume titled *Biosemiotics*. This volume was predicated on a book they edited in 1980, *Speaking of Apes,* that presented a detailed critical evaluation of current investigations of the ability of apes to learn language. Sebeok showed in a profound critique of the way the experiments were constructed that it is very doubtful that apes have such capabilities. This work and its profound consequences are summed up and developed further in the book *Life Signs (2000)*. Thus biosemiotics does not entail that there are no significant differences between man and apes.

Sebeok argued that the biosphere and the semiosphere are linked in a closed cybernetic loop where meaning itself powers creation in self-excited circuits. This is a thinking that clearly encompasses the similar ideas as are considered in second-order cybernetics, autopoiesis and enaction theory (Varela).

Sebeok was famous for his unselfish supportive qualities and organizational abilities. He picked up manuscripts from talented and unknown scholars all over the world and helped to get them published. Now and then – as in the case of Jesper Hoffmeyer's *Signs of Meaning in the Universe* – he even devoted a whole issue of *Semiotica* to review and discuss this new work. Thereby he created the attention necessary to spread the message worldwide, which helped the scientist through an international reputation to secure his position in his own country and subject area. Sebeok was well aware of the problems of getting tenure and the dangers of loosing it, especially when engaging in the development of new areas within academia, particularly the interdisciplinary and transdisciplinary ones.

Sebeok's supportive work is thus behind many of the members of the biosemiotic community who are now members of the editorial board of this journal. In this way he nurtured the development of the biosemiotic field and its interaction with information theory, AI, cybernetics and autopoiesis, computer science, and information seeking. He also supported the publication of this editor's work on Cybersemiotics. Sebeok proposed for Hoffmeyer and Emmeche the guest editorship of a special *biosemiotica* issue of *Semiotica* [*127*(1/4),1999] to lay a publicly recognized international foundation for biosemiotics. He further gave Kalevi Kull the task of editing a whole, even larger, special volume on the significance of Jacob von Uexküll for biosemiotics [*Semiotica,134*(1-4), 2001]. Thus, Sebeok was very supportive of various schools of semiotics including the Copenhagen School of Biosemiotics as well as the Department of Semiotics at Tartu University in Estonia, as well as the development of a Cybersemiotics in the present journal.

As we see, a great deal of Sebeok's efforts were put into unselfish interdisciplinary and transdisciplinary organizational work. He had a modest and sensitive way to help people improve their thoughts and their work, and a unique way of establishing contacts between like-minded scholars from different corners of the world was essential to the establishment of this field. Sebeok was able to do this, not only because he regularly toured the world attending conferences and teaching, but because he was the "Mister E-mail" of semiotics. He was a master of using e-mail as an international immediate communication tool, answering everyone — sometimes within the minute. At the last Imatra meeting which he attended in 2000, where we celebrated his 80th birthday, he told me how he loved to get up early in the morning before anybody else was awake, to sit in his pajamas enjoying his e-mail contact with Europeans, who had just started work at that time. At that time of day you could have a nice chat with him.

Although he retired in 1991 he continued to give lectures and lecture-series on all continents, and continued to subtly pull strings, particularly of the semiotic network. He also led the *Semiotics Publications Department at Indiana University* almost until his death. I will not go into all his numerous medals, honors and honorary presidencies or his world wide lecture tours at universities and conferences, until his illness caught up with him and slowed him down for the last couple of years of his life. He produced more than 600 books and articles. You will find the most relevant in the reference lists of the articles in this issue.

It is no wonder that Sebeok earned the nickname "Mister Semiotics." However, many of us will first of all think of him as "Mister Biosemiotics." Based on his hard work and in his spirit the third Biosemiotic Gathering will be held in Copenhagen July 11-14. 2003.

ASC
American Society for Cybernetics
a society for the art and
science of human understanding

The Integration of Second Order Cybernetics, Cognitive Biology (Autopoiesis), and Biosemiotics

Søren Brier, Trustee

In my view there are some very interesting commonalities between second order cybernetics, cognitive biology as based on autopoiesis, and Sebeok's Peircean semiotics. But there are also some interesting differences that lead me to the conclusion that they need each other. As an important part of biosemiotics rests on Sebeok's interpretation of Jakob von Uexküll (as the founding father of biosemiotics) I will refer to Uexküll's cybernetic model. For those that are not familiar with his functional cycle of perception and cognition I refer to Figure 1 below. The cybernetic inspiration is quite obvious, and the 'object' is constructed for 'the inner world of the subject' by a functional feedback circle between perceptual and effector cues. This is quite close to von Foerster's idea of 'cognitive eigen values' as I shall discuss below.

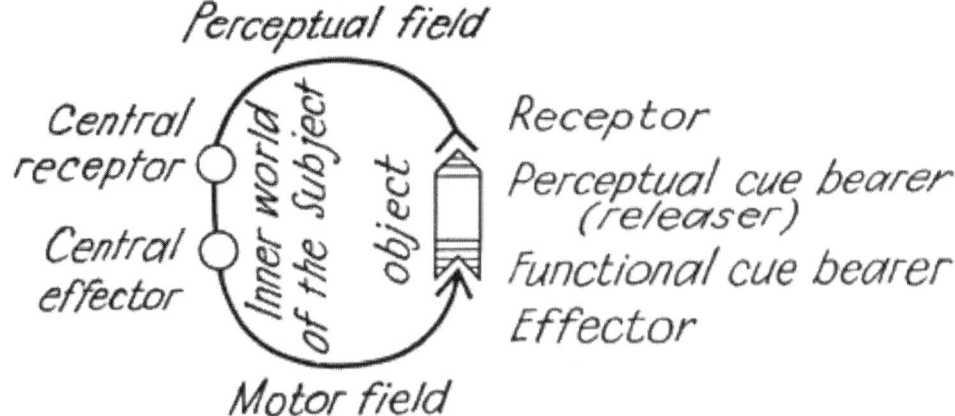

Figure 1: Functional Cycle

Further I have been most interested in Luhmann's further use of Maturana and Varela's autopoiesis theory to construct a general theory of social communication. But since the autopoiesis concept in biology is the crucial turning point I will mostly discuss Maturana's understanding of that. But let me consider the similarities between these perspectives first:

1. As second order cybernetics takes cybernetics and systems research to a new level by including the observer biosemiotics takes semiotics to a new level by including all living systems in semiosis.
2. In both cases this new level is achieved through a bio-constructivism, where all living systems are seen as constructing their own "life-world." In biosemiotics it is often called 'Umwelt' after von Uexüll. Maturana speaks of the organism's 'cognitive domain'. Von Foerster sees a cognitive world constructed of 'eigen values' of the nervous system's cognitive processes. Eigen values are stable systems of recursive processing that stabilize in the mind and enable us to (re)cognize things.
3. In all these systems of thought the bio-constructivism leads to an idea of 'closure'. The term is mostly used in connection with autopoiesis, but both von Foerster and von Uexküll have clear indications of that the 'life world' or 'signification sphere', as I call it, is all there is for the organism.
4. In all these systems there is **no** stream of 'information' from the environment going directly into the cognitive system of the organism that is picked up and gives a more or less 'objective' picture of the "real environment."
5. All acknowledge that "reality" or "the environment" exists as some kind of limit that puts a 'constraint' on the possible ways an organism can exist as an organism. Von Foerster is most explicit about accepting that the environment has to have energy and structure, von Uexküll also seems to accept some kind of real world outside the many Umwelts, as he calls these 'subjective worlds'.
6. They all agree that life and cognition are aspects of the same thing. Peirce and Sebeok use the term 'semiosis' and 'signification' for cognition. But broadly speaking they are saying that life, cognition and communication coincide,
7. Maturana, von Uexküll and von Foerster all discuss the kinds of experience that can arise under various circumstance; and, all use examples that have to do with vision. (What is it to see? What the frogs brain tell the frogs mind. Through the eyes of the others. ...and so on.) However none of them provide an explicit theory of the organism's first person phenomenological experience and the difference between experience and what goes on in the nervous system.
8. In biosemiotics Uexküll's stationary world view is transferred to Peirce's evolutionary world view. Through this operation biosemiotics, second-order cybernetics and autopoiesis share the evolutionary constructivist view of the origin of organisms, their cognition and ecological 'niches'.
9. None of them consider organism as deterministic machines. But cybernetics and autopoiesis are both more machine-like in their language than the biosemioticians. Von Foerster for instance refers to living systems, including humans as non-trivial machines.

10. Although von Uexküll clearly has a phenomenological view of the organism neither he, von Foerster, Maturana nor Sebeok have 'a theory of mind' or how first person experiences appears in a physical world. However, Peirce provides such a view which is only beginning to be discussed in semiotics.
11. All three views more or less explicitly take life as a basic or constituent aspect of reality, and not something invented by chance out of a physical deterministic world. As we shall see, Peircean biosemiotics differs from the others as Peirce's explicit metaphysics supports this stance.
12. As I pointed out in my ASC-column in our issue in honor of Varela (9(2),77-82), the development of second order cybernetics and biological cognition that Varela accomplished in his 'Calculus for self-reference' brings the foundation very close to Peirce's triadic relational category theory. This is an important part of the metaphysical theory behind Peirce's triadic theory of semiosis.

Now, to the interesting differences that I see among these views, which in my opinion make the construction of a Cybersemiotics both necessary and valuable:

1. The concept of 'structural coupling' is unique to autopoiesis, although von Foerster's concept of 'things as cognitive eigenvalues' is close to that, and von Uexküll has a more vague idea of the same. Structural coupling seems to be the prerequisite for generating cognitive eigen values which make cognitive objects possible. Structural coupling is necessary for the sudden construction of patterns that attain meaningfulness in the perceptual field, such as the 'sign stimuli' in the ethological paradigm of animal cognition, communication and behavior.
2. Maturana and Varela point out that it is the autopoietic character of living systems that makes it possible for them to conserve structural couplings. Through these structural couplings it is possible to establish von Foerster's eigenvalues of cognition. I suggest that this is what Peirce called the Interpretant that is the sign in our mind that makes us see/recognize something as a thing. Peircean biosemiotic build on Peirce's unique triadic concept of semiosis, where the 'interpretant' is the sign concept in the organism's mind that is its interpretation of what the outer sign vehicle "stands for," for instance, that a raised fist is at 'threat'. This is, of course, quite contrary to what Maturana proposes; i.e. no internal "representation" as such, rather a continuous flow of configurations within the nervous system, in a sensory motor closed loop, in which some configurations become more likely and appear as regularities. According to Maturana a nervous system is a detector of configurations within itself – and these do not take on the "solidity" of the "objectness" of interpretants.
3. Peirce's differentiation between the immediate object of semiosis and the dynamic object that is all we in time can get to know about it is an evolutionary solution to the problem of the relation between the significations sphere or "life world" of the organism and 'the environment or universe' outside it. This view is part of biosemiotics.

4. Peircean biosemiotics is based on Peirce's theory of mind as a basic part of reality (in Firstness) existing in the material aspect of reality (in secondness) as the 'inner aspect of matter' (a view called 'hylozoism') manifesting itself as awareness and experience in animals and finally as consciousness in humans. Combining this with a general systems theory of emergence, self-organization and closure/autopoiesis it constituted an explicit theory of how the inner world of organism is constituted and therefore how first persons views are possible and as real as matter.
5. Through this foundation for semiosis a theory of meaning and interpretation including mind – at least as immanent inside nature – is possible and cybernetic views of information, autopoietic views on languaging can be combined with pragmatic theories of language in the biosemiotic perspective (as I am offering models of in forthcoming papers).

This is why I find Sebeok's work on constructing a Peircean biosemiotics so important for much of the work published in this journal, in particular for that work which encompasses second order cybernetics and autopoiesis. Biosemiotics makes it possible for us to add a theory of mind, meaning and signification to our views on cognition. My version of this is what I call *Cybersemiotics*; others may develop different conceptions.

The discussion in this column is partly based on the following papers:

Brier, S. (1996): From second-order cybernetics to cybersemiotics: A semiotic re-entry into the second-order cybernetics of Heinz von Foerster, *Systems Research*, *13*(3), 229-244.
Brier, S. (1997): What is a possible ontological and epistemological framework for a true universal 'Information Science': The suggestion of cybersemiotics. *World Futures*, *49*, 287-308.
Brier, S. (2001): Cybersemiotics and Umweltslehre. *Semiotica*, *134*(1/4), 779-814. (Special issue on Jakob von Uexküll)

Signs of Life:
A Review of *I segni della vita. La semiotica globale di Thomas A. Sebeok*[1]

Marcel Danesi [2]

As semiotics enters the twenty-first century, the work of one of the greatest semioticians of all time, the late Thomas A. Sebeok (1920-2001), is finally being appreciated as providing a map of where the field has been and a blueprint of where it should be going. Known as the biosemiotic or global semiotic movement, the Sebeokian agenda is starting to attract an increasing larger cadre of semioticians and scholars in cognate disciples. This is the main point emphasized by Augusto Ponzio and Susan Petrilli of the University of Bari in a book that is much more than just a laudatory assessment of Sebeok's work, but also a solid contribution to the global semiotic movement itself. In this way, the two scholars chart the course that the study of semiosis will need to pursue in the future. The intent of their volume is thus: (1) to introduce or reintroduce Thomas A. Sebeok to an audience of researchers and students who might be interested in finding out what the global study of semiosis entails through an in-depth analysis of his pioneering works (e.g. Sebeok 1976, 1979, 1981, 1986, 1991a, 1991b, 1994, 2001, Sebeok and Danesi 2000); and (2) to provide a theoretical discussion of the basic notions in biosemiotics, the discipline that seeks to understand semiosis and communication across species.

The basic message in the Sebeokian paradigm is that life is semiosis—the innate faculty for producing and understanding signs. The signs that we produce are thus hardly arbitrarily-conceived artifacts, but rather they are structures that emanate from bodily experiences. They are "signs of life," so to say. People and animals are so vastly different, yet, paradoxically, very much the same. That is the contradiction of life. No life science other than biosemiotics is capable of penetrating this contradiction, as Ponzio and Petrilli so cogently argue.

The crucial distinction that Sebeok made in his writings was between semiosis as a product of biological processes, and representation as the activity of capturing, portraying, simulating, or relaying impressions, sensations, perceptions, ideas, etc. in conventionalized ways—through language, art, music, etc. Semiosis is a product of Nature, representation of Culture. Representational strategies allow human beings to refer to virtually anything they notice or find interesting in their world. Indeed, representation is so powerful cognitively, that human beings rarely fail to differentiate

---

1. By Augusto Ponzio and Susan Petrilli. Milano: Spirali, 2002.
2. Program in Semiotics and Communication Theory, Victoria College, University of Toronto, Toronto, Ontario M5S 1K7, Canada. E-mail: marcel.danesi@utoronto.ca. This paper is dedicated to the memory of Thomas A. Sebeok (1920-2001).

themselves from their representational forms. In chapter 3 of *I Think I Am a Verb* (1986), Sebeok brings this out by explaining the title of his book, which refers to a phrase uttered by the eighteenth President of the United States, Ulysses S. Grant, just before he died. For Sebeok it encapsulates the fact that human beings see themselves as signs. Verbs refer to actions, change, existence; they are perfect instantiations of sign activity, of the infinite capacity to generate signification—literally the "making of signs"—in order to construct models of reality. Grant's phrase thus nicely captures the essence of the human condition—the urgent need to capture, represent, and interpret the dynamic flux of the world in the form of signs, including the mystery of the Self. Sebeok's work examine the dynamics and manifestations of this very condition—which Ponzio and Petrilli call "dialogics"—an internal dialogue that constitutes the construction of the Self through signification, which then manifests itself in acts of true dialogue among interlocutors.

The Ponzio and Petrilli book is thus an important interpretive filter through which the whole biosemiotic movement can be viewed. It constitutes a wonderful engagement with the biosemiotic agenda, thus acknowledging Sebeok's legacy in a concrete way. In my view, there is no more appropriate way to pay homage to anyone than to focus attention on the relevance of his ideas. The book thus has both theoretical and practical value. It can be used as a basis for making nonverbal semiosis the starting point for understanding verbal semiosis and semiosis in general. It can also be used a complementary text in advanced university courses in semiotics. Unfortunately, it is written in Italian, and thus limits its use to those who know that language. A translation in other languages is something that would be welcome, since the writing is simultaneously explanatory and thought-provoking; usable by student and scholar alike.

References

Sebeok, T. A. (1976). *Contributions to the doctrine of signs.* Lanham: University of America Press.
Sebeok, T. A. (1979). *The sign and its masters.* Austin: University of Texas Press.
Sebeok, T. A. (1981). *The play of musement.* Bloomington: Indiana University Press.
Sebeok, T. A. (1986). *I think I am a verb: More contributions to the doctrine of signs.* New York: Plenum.
Sebeok, T. A. (1991a). *A sign is just a sign.* Bloomington: Indiana University Press.
Sebeok, T. A. (1991b). *Semiotics in the United States.* Bloomington: Indiana University Press.
Sebeok, T. A. (1994). *Signs: An introduction to semiotics.* Toronto: University of Toronto Press.
Sebeok, T. A. (2001). *Global semiotics.* Bloomington: Indiana University Press.
Sebeok, T. A. and Danesi, M. (2000). *The forms of meaning: Modeling systems theory and semiotics.* Berlin: Mouton de Gruyter.

Call for Papers
First Annual
Systems Theory Conference
May 23-25 2003

The Center for Corporate Communication (CCC) at the Copenhagen Business School invites researchers to submit abstracts toward a conference about the systems theory of Niklas Luhmann, its relationship to other theories, and its usefulness with regard to empirical work..

The conference has the twofold purpose of testing systems theory in relation to other theories and in relation to empirical research. Although Luhmann tried to position systems theory and demonstrate its productiveness in relation to these fields, many unsolved issues remain.

The crucial point is that systems theory be the point of departure for a comparison of theoretical scope, consistency and productiveness. The theoretical center is the question of second-order observation.

Keynote speakers :

· Dirk Baecker · Hans Ulrich Gumbrecht · Urs Stäheli · Niels Åkerstrøm Andersen

Submissions could be in the following categories:

· Discourse theory · Systems theory in alternative forms · Structuration theory · Semiotics · Communicative action theory · Constructivism · Deconstruction · Narrative theory · Actor-network theory

It is also possible to include particular themes or key concepts such as paradoxy, action, individual, field, etc., and compare their interpretation within different theories. In order to avoid an unproductive dispersal it is a requirement that this happens based on Luhmann's systems theory.

The conference will take place in Copenhagen May 23-25 2003. Abstracts of approximately 400 words should be sent to Center for Corporate Communication, by the end of February, 2003.

ccc@cbs.dk

To recieve the brochure, please send a postal adress. For further details visit:

http://asp.cbs.dk/ccc/luhmann.